Semiconductor Devices: testing and evaluation

BY THE SAME AUTHOR:

Reliable Electronic Assembly Production

Materials in Electronics

Electronic Engineering Processes

Electronics and Environments

Semiconductor Devices: testing and evaluation

C. E. JOWETT

London
BUSINESS BOOKS LIMITED

First published 1974

© CHARLES ERIC JOWETT 1974

All rights reserved. Except for normal review purposes, no part of this book may be reproduced or utilised in any form or by any means, electronic or mechanical, including photocopying, recording, or by any information storage and retrieval system, without permission of the publishers.

ISBN 0 220 66221 5

This book has been set in 10 on 12 Baskerville
and printed by REDWOOD BURN LIMITED
Trowbridge & Esher
for the publishers, Business Books Limited
(registered office: 110 Fleet Street, London EC4)
publishing offices: Mercury House, Waterloo Road, London SE1

MADE AND PRINTED IN GREAT BRITAIN

Contents

	Preface	x
Chapter 1	**INTRODUCTION**	1

General requirements — Test conditions — Permissible temperature variation in environmental chambers—Electrical test frequency—Accuracy—Accuracy of meters—Orientations—General precautions—Transients—Test conditions for electrical measurements—Pulse measurements — Soldering — Order of connection of leads — Reliability of testing—Loading effects—Multimeters—Digital voltmeters—Oscilloscopes—Reduction by provision of monitor points—Power supply problems—Supply arrangements—A.c. testing—Design for reliability—Step-stress testing as a design tool—Quantitative reliability—Accelerated life test techniques—Elimination of rogue failures—Need for a controlled environment

Chapter 2 **ENVIRONMENTAL TESTS** 15

Barometric pressure—Immersion—Insulation resistance—Moisture resistance — Steady-state operation — High-temperature life — Intermittent operational life — Salt atmosphere—Salt spray—Thermal shock—Temperature measurement: case and stud—Dew point

Chapter 3 **MECHANICAL TESTS** 33

Constant acceleration—Shock—Solderability—Soldering heat—Terminal strength—Stud torque—Lead fatigue—Bending stress—Vibration fatigue—Vibration noise—Vibration variable frequency—Vibration high frequency

Chapter 4 **SEMICONDUCTOR (TRANSISTOR) ELECTRICAL TESTS** 44

Collector-to-base breakdown voltage—Burnout by pulsing—Collector-to-emitter breakdown voltage—Drift—Floating potential—Breakdown voltage, emitter to base — Collector-to-base current — Collector-to-emitter voltage—Collector-to-emitter cut-off current—Collector-to-base voltage — Emitter-to-base current — Base-to-emitter voltage (saturated or non-saturated)—Saturation voltage and resistance—Forward current transfer ratio—Static input resistance—Static transconductance

vi   Semiconductor devices

Chapter 5   **MEASUREMENTS OF CIRCUIT PERFORMANCE AND THERMAL RESISTANCE**   57
Thermal resistance (collector cut-off current method)—Thermal resistance (forward voltage drop, emitter-to-base diode method—Thermal resistance (d.c. forward voltage drop, emitter-to-base continuous method)—Thermal response time—Thermal resistance (forward voltage drop, collector-to-base diode method)—Thermal time constant—Thermal resistance (general)—Thermal resistance (d.c. current gain, continuous method)

Chapter 6   **HIGH- AND LOW-FREQUENCY TESTS**   64
Small-signal short-circuit input impedance — Small-signal open-circuit admittance — Small-signal short-circuit forward-current transfer ratio—Small-signal open-circuit reverse-voltage transfer ratio—Small-signal short-circuit input admittance—Small-signal short-circuit forward-transfer admittance—Small-signal short-circuit reverse-transfer admittance—Small-signal short-circuit output admittance—Open-circuit output capacitance—Input capacitance—Direct inter-terminal capacitance—Diffusion capacitance—Depletion layer capacitance—Noise figure—Pulse response—Small-signal power gain—Extrapolated unity gain frequency—Real part of small-signal short-circuit input impedance—Small-signal short-circuit, forward-current transfer ratio cut-off frequency—Small-signal, short-circuit, forward-current transfer ratio

Chapter 7   **GENERAL DIODE TESTS**   81
Capacitance—D.c. output current—Forward current and forward voltage—Reverse current and reverse voltage—Breakdown voltage—Forward recovery time—Reverse recovery time—'Q' for variable capacitance diodes—Rectification efficiency—Average reverse current—Small-signal breakdown impedance—Surge current—Temperature coefficient of breakdown voltage—Small-signal forward impedance—Stored charge—Saturation current—Thermal resistance for signal diodes, rectifier diodes and controlled rectifiers

Chapter 8   **MICROWAVE DIODE TESTS**   95
Conversion loss—Detector power efficiency—Figure of merit — Intermediate frequency impedance — Output noise ratio—Video impedance—Burnout by repetitive pulsing—Burnout by single pulse

Chapter 9 **THYRISTORS** 105
Holding current—Forward leakage current—Reverse leakage current—Pulse response—Gate triggering signal—Instantaneous forward voltage drop—Rate of voltage rise

Chapter 10 **TUNNEL DIODE TESTS** 109
Junction capacitance—Static characteristics of tunnel diodes—Series inductance—Negative resistance—Series resistance—Switching time

Chapter 11 **FIELD-EFFECT TRANSISTORS** 114
Gate-to-source breakdown voltage — Gate-to-source voltage or current—Drain-to-source 'on' state voltage—Drain-to-source breakdown voltage — Gate reverse current—Drain current—Drain reverse current—Static drain-to-source 'on' state resistance—Small-signal drain-to-source 'on' state resistance test — Small-signal, common source, short-circuited input capacitance test—Small-signal, common source, short-circuit reverse-transfer capacitance—Small-signal, common-source, short-circuit output admittance and forward transadmittance—Small-signal, common-source, short-circuit input admittance—Note

Bibliography 125

Index 127

Preface

This book is intended to present uniform methods for testing semiconductor devices. It includes basic environmental tests to determine resistance to deleterious effects of normal elements and conditions surrounding normal operation together with physical and electrical tests.

For the purpose of this book, the term 'device' includes such items as transistors, diodes, voltage regulators, rectifiers and tunnel diodes. The test methods have been prepared to serve several purposes: to specify suitable conditions obtainable in the laboratory which give test results equivalent to conditions existing in use in an equipment and to obtain reproducibility of the results of tests. The tests described are not to be interpreted as an exact and conclusive representation of actual operation in any one geographical location, since it is known that the only true test for operation in a specific location, is an actual test at this point.

Rapid advances in device manufacture have led to comprehensive test requirements covering static characteristics, small-signal parameters and tests applicable to special types. The static characteristics are evaulated from d.c. measurements and form the main part of test requirements for all types of semiconductor device. For devices produced for special purposes, tests are designed to check their functioning in special modes, the most common set of tests being those which cover the switching characteristics of the device.

All these requirements have led to many new testing techniques.

The methods employed in this book have, as far as possible, been kept uniform in order to conserve equipment, man-hours and test facilities. In achieving this objective, it is necessary to make each of the general tests adaptable to a broad range of devices. The tests have been divided into four classes: environmental, mechanical, transistor and diode electrical tests.

To assist the reader of this book and as a reference, a system of symbols for device test parameters is now in common use and can be found in British Standard 3363.

Harpenden CHARLES E. JOWETT
January 1974

Chapter 1 Introduction

General requirements

Test conditions

All measurements and tests should be made at thermal equilibrium at an ambient temperature of 25°C and at ambient atmospheric pressure and relative humidity unless otherwise specified. Wherever these conditions are to be closely controlled in order to obtain reproducible results, the reference conditions should be as follows: temperature 25°C, relative humidity 50%, atmospheric pressure from 650 to 800 mmHg.

Permissible temperature variation in environmental chambers

When chambers are used, specimens under test should be located only within the working area. The chamber should be controlled and be capable of maintaining the temperature of any single reference point within the working area within ±2 degC. The chamber should be so constructed so that, at any given time, the temperature of any point within the working area will not deviate more than ±3 degC or ±3%, whichever is the greater, from the reference point, except in the immediate vicinity of specimens generating heat.

Electrical test frequency

The electrical test frequency should be 1 kHz ±25% except for power diodes where a frequency of 40–70 Hz should be used unless otherwise specified.

Accuracy

The absolute (true) values of the device parameters should be within the limits specified. Wherever possible, measurements and equipment calibration should be traceable to British standards.

Accuracy of meters

Accuracies of not less than 2% for d.c. measurements and 5% for a.c. measurements should be maintained on all ammeters and voltmeters.

2 Semiconductor devices

Orientations

X is the orientation of a device with the main axis of the device normal to the direction of the accelerating force, and the major cross-section parallel to the direction of the accelerating force.

Y is the orientation of a device with the main axis of the device parallel to the direction of the accelerating force and the principal base toward (Y_1) or away from (Y_2), the point of application of the accelerating force.

Z is the orientation of a device with the main and the major cross-sections of the device normal to the direction of the acceleration force.

For the case of configurations other than shown in Figs 1 and 2 the orientation of the device should be as specified.

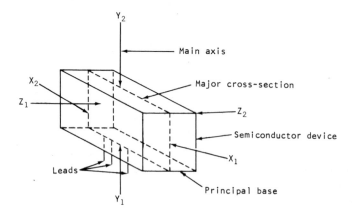

Figure 1 Orientation of a non-cylindrical semiconductor device to direction of accelerating force

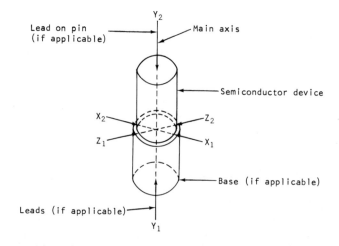

Figure 2 Orientation of cylindrical semiconductor device to direction of accelerating force

General precautions

The following should be observed in the testing of devices.

Transients

Devices must not be subjected to conditions in which transients cause a rating to be exceeded.

Test conditions for electrical measurements

Test conditions for all electrical measurements should be such that the maximum ratings are not exceeded. These precautions should include limits on maximum instantaneous currents and voltages. High series resistances (constant current supplies) and low capacitances are usually required. If low cut-off or reverse current devices are to be measured, for example, nanoampere units, care should be taken to ensure that parasitic circuit currents or external leakage currents are small, compared with the cut-off or reverse current of the device to be measured.

Pulse measurements

When the static or dynamic parameters of a device are measured under 'pulsed' conditions, in order to avoid measurement errors introduced by device heating during the measurement period the following should be covered in the detail specification:

1. Pulsed test should be placed by the test specified.
2. Unless otherwise stated, the pulse time (t_p) should be 250 to 350 μsec and the duty cycle should be 1–2%.

Soldering

Adequate precautions should be taken to avoid damage to the device during soldering required for test.

Order of connection of leads

Care should be taken when connecting a semiconductor device to a power supply. The common terminal should be connected first.

Reliability of testing

The reliability of an electronic product depends equally on the soundness of its design and on the use of reliable components and manufacturing methods. The evidence confirming that each of these factors has received sufficient attention in evolving the final product is obtained mainly from test results. The production departments regard the successful completion of over-all tests on the final equipment as a reasonable check that assembly has been correctly carried out and that devices used are within specification. The electronic designer satisfies himself that mean values and deviations of the test figures are within the limits expected from design calculations. So much depends on the correctness of these results that specially designed test equipments are usually provided even for small production batches, to guarantee as far as possible the proper execution of final tests.

Some aspects of performance, having a profound effect on reliability, are unfortunately difficult to check during production. This may be due to the following factors:

1 Inaccessibility of a part of the completed equipment, where long connecting leads would upset circuit operation.
2 The appropriate tests may require expensive test equipment only suitable for a research laboratory or only operable by a highly skilled engineer.
3 For tests at elevated or reduced temperatures, the testing time may be too long and expensive to consider carrying out on every piece of final equipment.

The solution invariably adopted is to perform these searching tests on a 'breadboard' or prototype model, before full production is commenced. These tests are supervised by the designer and generally include sufficient informative tests to reveal any shortcomings of his design. Satisfactory performance of the prototype under worst-case conditions of such variables as ambient temperature and supply variations is taken as evidence of correct design.

It is then assumed that, since the basic design is thus proved correct, the relatively simple final testing indicates correct manufacture. This assumption is not necessarily justified particularly in terms of performance at elevated temperatures. There are two unfavourable possibilities:

1 A circuit component has poor performance at high temperatures only; this will be undetected in final tests at room temperature and even though the electronic design was correct the tested unit will fail in service at high temperature.
2 The design has not allowed for the expected deterioration of a component at high temperature—the component used in the very thorough prototype tests may be better than its specification, so that many apparently satisfactory production units will fail in service.

The second possibility is alarming and indicates the need for careful design and thorough checking of at least two prototypes to reduce the probability of missing a design error of that type.

It is clear that the reliability of the final product depends greatly on correct design and that designers rely on the results of prototype tests to indicate unforeseen design errors. In spite of this, designers often take little interest in ensuring that these tests are properly conducted. Inexperienced engineers or technicians are expected to produce the results without falling into the traps avoided by the designer only after years of sad experience.

Although these are simple measurements in principle, there are several ways in which misleading results can occur.

Loading effects—multimeters

A typical multimeter has a resistance of 20 kΩ/V f.s.d. and is often used on a 10-V full-scale deflection range giving 200 kΩ of voltmeter loading. The obvious consequence is that the circuit is disturbed by this load and displays the lower voltage that results.

This principle is elementary, but wrong measurements often go unnoticed because the degree of error is generally not enough to cause alarm. Mistakes of this kind are increasingly common because transistors and FET's can operate successfully at low currents and can use load resistors of the same order as the multimeter.

For instance, the T_1 collector-earth potential in Fig. 3 is nominally +8 V, but would be measured 4.8 V on the 10 V range; the emitter should and does read about +5.3 V, because loading of 200 kΩ on the emitter has little effect owing to the low output resistance at the emitter. The base seems to be more negative than the emitter yet the emitter circuit must be passing current to reach any potential above earth! The use of a second multimeter and measuring both quantities simultaneously would show up the error but would still give the wrong base voltage.

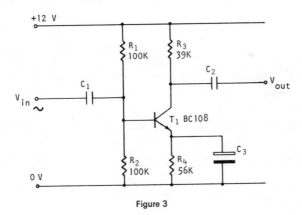

Figure 3

Using multimeters, the following should be routine: after each voltage measurement, the instrument should be switched to the next higher range and the measurement repeated. Any difference, other than reasonable reading errors, indicates that the loading is affecting the circuit.

Another problem in using a multimeter is the injection of mains or radio frequencies into the circuit under test or the provoking of self-oscillation in the circuit. Such signals can become rectified and thus change the d.c. levels being monitored.

The magnitude of this kind of signal and its effect are incalculable and the only safeguard is to monitor a suitable point in the circuit with an oscilloscope throughout the tests.

Current measurement is rarely attempted owing to the need for breaking connections to insert the meter. A common error when this is done is to forget the 0·5-V drop absorbed by most multimeters.

Loading effects—digital voltmeters

Owing to their very high input resistance and high accuracy, digital voltmeters are often used for critical d.c. measurements. The continuous monitoring of circuit behaviour by an oscilloscope is then particularly important because many digital voltmeters present a highly variable impedance, often being intermittently almost short-circuit. Some drive pulse trains into the circuit under test, while others have an extremely high parallel input capacitance (0.25 μF) on certain ranges, which could affect the normal operation of many circuits.

Loading effects—oscilloscopes

The oscilloscope is frequently used for d.c. measurements because it offers high input resistance and simultaneously monitors a.c. components, thus indicating unwanted pick-up and oscillation.

It is easy to overlook that the usual 1 MΩ input resistance can still load some circuits, e.g. that shown in Fig. 3, noticeably and even with the commonly attached 10:1 probe, the loading cannot always be ignored.

The input capacitance of a direct screened lead combined with that of the oscilloscope will often exceed 100 pF and oscillation of the circuit under test is more likely than with a multimeter using separated leads. The use of an attenuator probe having less than 10 pF capacitance at the cost of 10:1 loss of signal, is therefore, essential for many measurements.

In cases where the 7 pF of a typical probe still causes oscillation, a resistor of a few kilohms should be clipped to the probe and the remote end of the resistor used for connection to the circuits, as if it were the probe tip.

Loading effects—reduction by provision of monitor points

It is evident that loading effects can lead to wrong readings which are revealed as erroneous. Much can be done to avoid these errors by the provision of monitor points during the initial design. These are designed to minimise loading by the measuring equipment used during testing, and generally they provide the bonus that inadvertent short-circuiting of the measuring point to ground causes less damage to active devices. Figure 4 illustrates an audio circuit provided with perhaps an excess of monitor points.

Power supply problems

In the testing of new circuit designs, bench power supplies are commonly used and the circuits are tested using their built-in supplies only at a later stage. The reason is clear—until the whole equipment is fully designed, the

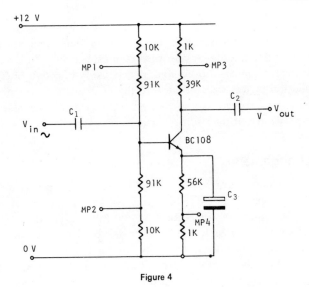

Figure 4

power supply load cannot be estimated and so its design is not begun until late in the development period. There are dangers in using bench supplies for the initial tests and some precautions are advisable.

Most d.c. supplies can be used for supplying current in one direction only. This applies to an unregulated supply derived from a.c. by diode rectification and also to a conventional stabilised d.c. supply using a series regulator, unless provided with a permanent 'bleed' load. If the circuit being driven uses two supplies of this type of the same polarity and part of the circuit uses the voltage between supplies, stability of one supply can fail. Figure 5 illustrates a possible configuration of this kind, where T_1 and T_2 form a multivibrator driving T_3. The current in the 10 V supply is reversed and a normal stabilised supply could lose stability. Note that when the same supplies are driving the whole equipment rather than the small part of it

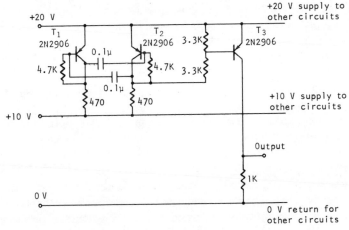

Figure 5

under test, it will function correctly provided the load has been correctly distributed. A cyclicly varying load current in the above situation may cause reverse current only at the peaks of each cycle. This often passes unnoticed since the mean supply voltage is scarcely affected, but may cause either oscillation or low gain because of the feedback thus introduced. The remedy for test purposes is to add bleed resistors to the supplies to ensure that the supply current does not reverse.

Supply arrangements

It is advisable when testing small parts of a large equipment to use individually variable supplies for each required voltage, rather than a specially made multiple supply as will be used in the final equipment. This enables the effect of variation of each line to be examined. Even if the test specification for the equipment does not require such a test to be carried out, it is useful to note the effect of, say, $\pm 10\%$ line variations. Naturally the circuit may fail to achieve its normal performance under these conditions, but complete failure to operate is usually a sign that the design is over-critical and warrants further investigation.

Some designs based on critical d.c. levels may have every right to fail, but the designer would be aware of this and accept this particular result. When a circuit requires several power supplies it is useful to add a master switch between the circuit and the supplies. This greatly increases the probability that the technician will take the trouble to switch off supplies before circuit modifications are made. It also ensures that variable controls on the supplies are not accidentally moved, in switching on or off. There is, however, a danger associated with the use of a master switch. In badly designed circuits that will fail catastrophically when some supplies appear before others, the master switch may approximate to simultaneous connections well enough not to cause damage. The circuit tests having been completed satisfactorily, this bad design feature will then appear as a production defect when using the normal multi-output supply. Additional tests should therefore be carried out to check the ability of the circuit to withstand all combinations of individual supply switching. Circuits unable to accept certain sequences of switch-on are unsatisfactory. Except in very rare cases it is much better practice to redesign the circuit than to design a specially interlocked power supply.

Most modern bench power supplies are provided with a current-limiting protection circuit.

When a load takes current in excess of a pre-set level, normal stabilisation ceases, the supply becomes a constant current source and its terminal voltage falls until the overload is removed. A lamp indication is usually given when an overload is present.

Although such a protection arrangement is desirable, especially when testing newly constructed circuits possibly containing wiring errors, false circuit operation can result. This is a hazard whenever pulsating load currents exist, as in class A-B, B or C amplifiers or oscillators. If the peak current exceeds the trip level, the supply voltage falls until the current falls and affects current operation. This may occur for only a small proportion of each cycle so that the overload indicator lamp remains unlit.

In circuits carrying intermittently large currents it is therefore good practice to monitor the supply lines, using an oscilloscope. Overloading is indicated by a drop in supply voltage for the duration of the excess current. When the equipment under test is provided with its own rectifier/stabiliser circuits to be driven directly from the mains, the effect of mains variation must be checked. The usual method is to insert a continuously variable transformer between the mains and the unit and to record the unit's performance at the specified extremes of input voltage.

Unless the variable transformer is rated at many times the power actually consumed by the circuit, it is likely that the input waveform to the unit will be heavily distorted when driving a peak rectifier system. The output impedance of such a transformer is noticeably higher than that of a similarly rated fixed type and the current pulses demanded by the diodes cause large voltage drops.

Since it is the voltage during this time that determines the rectified output, the stabiliser will operate as if the input were less than an average or r.m.s. reading of the a.c. input shows, giving misleading results. It is, therefore, important to monitor the alternating input voltage, using an oscilloscope, when such a test is in progress. Appreciable flattening of the waveform peaks shows that the transformer impedance is excessive. Either a higher-wattage device, or a tapped, fixed transformer should be used to overcome this problem.

A.c. testing

Many of the pitfalls described above are equally applicable to a.c. tests but are less likely to cause confusion. This is because most of the difficulties pointed out were due to the test equipment interfering with correct circuit operation. The engineer making a.c. measurements is almost certain to notice such abnormalities as spurious oscillation or distortion when examining these waveforms for the purpose of his measurement. When measuring 'only d.c.' this may not be the case.

There are, however, some errors caused by test gear which do not attract attention due to obvious waveform changes. For instance the oscilloscope has other peculiarities as well as the loading effects. The usual 10:1 attenuating probe, often used because of its small loading effect, is supplied with a definite lead length. In many well known probes this is not a length of normal coaxial cable but one with a resistive inner conductor. It is designed for use with the oscilloscope; neither the shortening of this lead or the addition of an extension lead of coaxial cable between the lead and the oscilloscope is permissible. The result would be an uneven frequency response giving distortion of pulse waveforms and incorrect amplitude of sine wave signals.

Many oscilloscopes have a twin-beam or multiple-beam display. If this is achieved by using one time-base to drive nominally identical guns, then the time relationship between signals applied to the beams is correctly reproduced (except for geometric errors in the guns). It is common practice however to use a single gun tube and to make its beam switch between the two signals, giving the appearance and facilities of a twin-beam display. At high time-base speeds the switching is performed at the end of each sweep to avoid the need for the very fast switching that would be needed during the sweep to avoid a dotted-line display.

The difficulty now appears that, if each signal produces a trigger pulse to initiate its own sweep, relative time synchronism is lost. Thus if two waveforms identical in shape but displaced in time, i.e. phase, are applied to the two beams respectively, they appear to be identical in time as well as shape.

It is important, then, when using the 'alternate sweep' system to trigger both sweeps from the one signal, using the 'External trigger' input.

Another snag is associated with this situation; the external trigger input of many oscilloscopes has a very low input resistance and the use of a 10:1 probe just for triggering purposes may be necessary.

When measuring the amplitudes of a.c. signals the oscilloscope may not be sufficiently accurate. Analogue voltmeters often give accuracies of the order of 1% for measurements in the audio region, and digital voltmeters equipped with a.c. adapters achieve even better accuracy. Most oscilloscopes are specified to about 3% and are also more tiring to read if a succession of measurements is to be made.

There is, therefore, good reason to use a meter instrument but great care needs to be taken to ensure that the correct measurement is really being made by the instrument. Most are basically rectified-average instruments; some (especially r.f. meters) are peak or peak-to-peak; a few are 'true r.m.s.', i.e. they read r.m.s. regardless of waveform. Dials are generally scaled in r.m.s., valid—except for the 'true r.m.s.' type—only if the waveform is distortionless sine wave.

Many meters, especially digital voltmeters, cause severe and non-linear circuit loading especially at frequencies near the response limit of the instrument. After deciding that a meter instrument is suitable, it is therefore particularly important to monitor the signal with an oscilloscope throughout the measurements in order to view the wave-shape.

Testing is at least as important as any other phase in the evaluation of a piece of electronic equipment. Care must be taken that the test gear does not noticeably change circuit operation, that the instruments are used so as to give correct results and that the appropriate interpretation is given to the results in view of the limitations of the test equipment.

Design for reliability

The need for reliability as a deliberate aim (as opposed to being incidental to good design) has been increasingly realised in recent years. Electronic systems of increasing complexity use many thousands of devices and a seemingly insignificant failure rate occurring in a large device population can result in a high fault rate in the system. Add to this the increasing tendency to rely on electronics where servicing is impossible or extremely expensive (satellites, missiles in flight, submerged or buried repeaters) or where human safety is involved (marine navigation, nuclear control, blind landing) or where faults mean loss of revenue (commercial television, process control computers, public telephone systems) and the case for reliable devices is established.

How is reliability to be achieved? Firstly, by designing the product to eliminate weaknesses; secondly, by manufacturing it under close control; and thirdly, by a continuing process of observation to ensure that the requirements are being met consistently and that adverse trends are corrected.

By environmental testing we mean the subjection of device material and equipment to carefully controlled conditions in respect to one or more of the environments in which it is expected to operate, for example, steady temperature, humidity, vibration, contamination, thermal cycling, corrosive atmospheres, pressure, voltage.

The designer has become accustomed to using the climatic test as a check on the suitability of materials and finishes, the vibration test as a check on structural rigidity, mechanical or thermal shock as a check on stability; he now knows that such tests can give him a rapid assessment of the probable behaviour of his product in service, and he uses them as a design tool.

Suitable programmes of environmental testing, usually of relatively short duration, have customarily been used for design acceptance and qualification approval testing. They indicate the ability of a given type of device to survive under the chosen conditions, but in general have not been able to provide useful data on reliability.

The extension of the technique of testing under controlled environment to provide more specific information on reliability will now be outlined.

In the design of equipment, reliability engineers may be attached to a project team to review and modify design and construction, so eliminating weaknesses without altering the system conceived by the original designers.

The design of devices is a more intimate and integrated subject which does not lend itself to this approach. The causes of failure are often inherent in the materials, processes and structures of the device, changes to which may fundamentally alter its characteristics. The designer must also be a reliability engineer, and reliability must become a basic design objective, rather than a feature which is grafted on after the design is established.

Reliability in design comes from an accurate knowledge of the causes of failure, and the adoption of techniques to eliminate them. Environmental testing plays an essential part in this process, but success in improving the reliability of a device brings a penalty as well as a reward.

As the life and survival rate of the device improves, normal environmental testing becomes progressively less useful, because of the extended testing periods necessary to reveal the remaining weaknesses which will cause failure during life.

It therefore becomes necessary to adopt accelerated test techniques, in which the devices are subjected to a selected and carefully controlled environment, and are deliberately overstressed in order to induce failures at a faster rate than normal. Such methods are capable of providing valuable information in a relatively short time, but can also be misleading if applied without proper consideration of the factors involved.

Step-stress testing as a design tool

Step-stress testing has been proposed as a method of obtaining short-term life ratings under extreme conditions of environment. Similar methods can be applied to the study of failure mechanisms and as a means of interpreting accelerated life tests to provide predictions of life under normal environments.

Correct application depends on a proper analysis of the failure mechanisms to which the device is subject. It is then necessary to choose a suitable means of stressing which will accelerate the failure mechanism under examination without introducing extraneous causes of failure. Devices are then subjected to test periods at several different elevated stress levels, and the resultant failures are recorded.

Failures include not only catastrophic failures (such as short-circuit, open-circuit, physical destruction) but also parameter failures resulting from the drift of a specified characteristic outside specified limits. By suitable plotting of failures against time and stress level, it is possible to estimate the lifetime which can be expected at normal stress levels. The most commonly used accelerating stress is temperature, because certain failure mechanisms are found to be dependent on temperature according to the function:

$$t = A\exp[b/T]$$

where t is the lifetime to a given percentage of total failures, T the temperature during life (in kelvins), b a constant dependent on the activation energy of the deterioration process and A a constant dependent on the rate of deterioration. This relationship can also be expressed:

$$\log_e t \propto 1/T$$

By carrying out several accelerated life tests at different elevated temperatures the results can conveniently be plotted and extrapolated to give a prediction of life expectancy at any desired temperature within the normal operating range.

Among other types of stress which may be adopted, according to the nature of the failure mechanism, are voltage, power dissipation, humidity, pressure and vibration.

The over-stress method is particularly applicable to simple basic devices in which systematic failure mechanisms are clearly definable. The method becomes progressively less attractive as the complexity of the device increases because each of several causes of failure has to be investigated separately, and allowance made for possible interaction between them.

The over-stress technique can however be of considerable assistance to the designer wishing to investigate and eliminate certain types of failure mechanism, and enables him to predict with reasonable confidence the effect of his work on the reliability of the product.

Quantitative reliability

The classical method of measuring reliability requires a number of devices to be tested for an extended period, failures being observed and used to estimate a failure rate, which is usually expressed as a percentage per thousand hours. In order to obtain reasonable confidence in this estimate a large number of test-hours must be completed, particularly if the reliability of the device under test is good.

This method is only applicable to devices which are made on a large scale by methods which are not subject to short-term variations. Having ascribed a value to the device failure rate, cumulative testing of batches from subsequent production is necessary to maintain confidence again to a level determined by the scale of sampling adopted. In practice the results obtained are not particularly accurate; for economic reasons, predictions are usually based on small numbers of failures in a given batch.

Accelerated life test techniques

The use of step-stress tests has already been mentioned as a tool for the designer. This method can also be used to provide extrapolation factors for accelerated testing of production samples. Cumulative reliability testing can be carried out more economically, and the results of individual tests can be obtained in time to take corrective action if unfavourable trends are observed.

The method is still subject to the limitation that the particular environment chosen to provide accelerated results may not accelerate all failure mechanisms equally. It is therefore important to understand the physical or chemical processes which cause failures in the device, and to apply the right criteria to defining what is meant by a failure; parametric drift which leads to equipment malfunction is as significant as catastrophic failure, and is more frequently encountered.

Many devices exhibit systematic drift of parameters, and change in characteristics beyond a specified level will determine useful life. The lifetime may be shortened by voltage, temperature or humidity, or by a combination of one of these with vibration or shock. The correct accelerating stress must be chosen for each type after a thorough study of failure mechanisms.

Elimination of rogue failures

In studying the failure pattern of a device, it is important to distinguish between different types of failure. Rogue failures result from lack of control in manufacture and produce failures which are not necessarily consistent from batch to batch.

Systematic failures, on the other hand, occur at a rate determined by the time-dependent physical changes resulting from use (or storage) of the device. Variations in the measured failure rates will occur from batch to batch, even for systematic failures, but the results of a series of tests will conform to a pattern of distribution.

Rogue failures, besides being undesirable in themselves, prevent effective analysis of systematic failure rates and may invalidate reliability predictions. This is true for failure rates based on small numbers of defectives, and also when accelerated testing is employed.

Tests have been introduced for some types of device in an attempt to eliminate rogues. Usually these assume that such failures will occur relatively early in life and that by consuming a small fraction of the device's useful life by burn-in-tests, the rogues will fail before leaving the manufacturer.

In certain cases, the normal ratings of the device may be exceeded in order to reduce the testing time. In such cases it has to be proved that such tests do not introduce additional potential failures, and that the mean lifetime is not significantly shortened.

An extension of non-destructive testing techniques offers the possibility of further improvements in device reliability. New methods are now being investigated with the aim of predicting failures before they occur. New types of test equipment can permit critical examination on the production lines of the construction and performance of devices in ways which hitherto could only be done in the laboratory.

Among the methods now being adopted as production tools are X-ray viewers, infra-red scanners, mass spectrometer gas detectors, low-level contact resistance monitors and noise measuring sets.

By subjecting the device to a suitable environment, whilst under examination, or by means of a preconditioning cycle before examination, it will be possible to detect and eliminate many rogues.

Need for a controlled environment

Accuracy in the results obtained depends in large measure on strict control of the environmental conditions applicable to the accelerated tests. Such precautions are also necessary to preserve correlation of the results of step-stress tests, or errors may occur in the calculation of acceleration factors. Experience has shown that an uncontrolled and unobserved variable in the over-all environment can destroy the correlation between test results and cause incorrect predictions. For this reason, these tests should be carried out by those with experience in testing under controlled environments.

Chapter 2 Environmental tests

Barometric pressure, reduced (altitude operation)

The barometric-pressure test is performed under conditions simulating the low atmospheric pressure encountered in the non-pressurised portions of aircraft and other vehicles in high-altitude flight. The test is intended primarily to determine the ability of component parts and materials to avoid dielectric-withstanding-voltage failures due to the lowered insulating strength of air and other insulating materials at reduced pressures. Even when low pressures do not produce complete electrical breakdown, corona and its undesirable effects, including losses and ionisation, are intensified. Low barometric pressures also serve to decrease the life of electrical contacts, since intensity of arcing is increased under these circumstances. For this reason, endurance tests of electromechanical components are sometimes conducted at reduced pressures. Low-pressure tests are also performed to determine the ability of seals in components to withstand rupture due to the considerable pressure differentials which may be developed under these conditions. The simulated high-altitude conditions can also be used in the investigation of the influence on component operating characteristics of other effects of reduced pressure, including changes in the permittivity of materials, reduced mechanical loading on vibrating elements (such as crystals) and decreased ability of thinner air to transfer heat away from heat-producing components.

Procedure

The device is to be tested in accordance with the following procedure. During this test and for a period of 20 min before, the test temperature should be $25 \pm 3°C$. The device should have the specified voltage applied and monitored over the range from atmospheric pressure at the specified minimum pressure and return for any device malfunctions. A device which exhibits arc-overs, harmful coronas, or any other defect or deterioration which may interfere with the operation of the device, is to be considered a failure.

Apparatus

The apparatus used for the barometric-pressure test consists of a vacuum pump and a suitable sealed chamber having means for visual observation when necessary of the specimen under test. A suitable pressure indicator should be used for measuring the simulated altitude (in feet or metres) in the sealed chamber. The specimens are mounted in the test chamber and the pressure reduced to the value indicated in one of the following test conditions, as specified. While the specimens are maintained at the specified pressure, and after sufficient time has been allowed for all entrapped air in the chamber to escape, the specimens will be subjected to the specified tests (see Table 1).

Table 1

Test condition	Pressure maximum inHg	mmHg	Altitude ft	m
A	8.88	226.00	30,000	9,144
B	3.44	87.00	50,000	15,240
C	1.31	33.00	70,000	21,336
D	0.315	8.00	100,000	30,480
E	0.043	1.09	150,000	45,720
F	17.3	439.00	15,000	4,572
G	9.436×10^{-8}	2.40×10^{-6}	656,000	200,000

Details

The following details should be recorded:

1. Method of mounting.
2. Test-condition letter.
3. Tests during subjection to reduced pressure.
4. Tests after subjection to reduced pressure if applicable.
5. Exposure time prior to measurements, if applicable.

Immersion test

This test is performed to determine the effectiveness of the seal of component parts. The immersion of the part under evaluation into liquid at widely different temperatures subjects it to thermal and mechanical stresses which will readily detect a defective terminal assembly, or a partially closed seam or moulded enclosure. Defects of these types can result from faulty construction or from mechanical damage such as might be produced during physical or environmental tests. The immersion test is generally performed immediately following such tests because the tests will tend to aggravate any incipient defects in seals, seams and bushings which might otherwise escape notice. This test is essentially a laboratory test condition, and the procedure is intended only as a measurement of the effectiveness of the seal following this test. The choice of fresh or salt water as a test liquid is dependent on the nature of the component part under test. When electrical measurements are made after immersion cycling to obtain evidence of leakage through seals, the use of a salt solution instead of fresh water will facilitate detection of moisture penetration.

This test provides a simple and ready means of detection of the migration of liquids. Effects noted can include lowered insulation resistance, corrosion of internal parts and appearance of salt crystals. The test is not indended as a thermal-shock or corrosion test, although it may incidentally reveal inadequacies in these respects.

Procedure

This test consists of successive cycles of immersions, each cycle consisting of immersion in a hot bath of fresh (tap) water at a temperature of $65(+5, -0)$°C, followed by immersion in a cold bath. The number of cycles, duration of each immersion, and the nature and temperature of the cold bath should be as indicated in the test condition listed in Table 2.

Table 2 Immersion test conditions

Test condition	Number of cycles	Duration of each immersion, min	Immersion bath (cold)	Temperature of cold bath, °C
A	2	15	Fresh (tap) water	25 (+10 −5)
B	2	15	Saturated solution of sodium chloride and water	25 (+10 −5)
C	5	60	Saturated solution of sodium chloride and water	0 (±3)

The transfer of specimens from one bath to another must be accomplished as rapidly as practicable. After completion of the final cycle, specimens shall be thoroughly and quickly washed and all surfaces wiped or air-blasted clean and dry.

Measurements

Unless otherwise specified, measurements are to be made at least 4 hr, but not more than 24 hr, after completion of the final cycle. Measurements to be made as specified.

Details

The following details must be recorded:

1. Test condition letter.
2. Time after final cycle allowed for measurements, if other than specified.
3. Measurements after final cycle.

Insulation resistance

The intended purpose of this test is to measure the resistance offered by the insulating members of a component part to an impressed direct voltage tending to produce a leakage of current through or on the surface of these members. A knowledge of insulation resistances is important, even when the values may be limiting factors in the design of high-impedance circuits. Low insulation resistances, by permitting the flow of large leakage currents, can disturb the operation of circuits intended to be isolated, for example, by forming feedback loops. Excessive leakage currents can eventually lead to deterioration of the insulation by heating or by direct-current electrolysis. Insulation-resistance measurements should not be considered the equivalent of dielectric-withstanding voltage or electric breakdown tests. A clean, dry insulation may have a high insulation resistance, and yet possess a mechanical fault that would cause failure in the dielectric-withstanding voltage test. Conversely, a dirty, deteriorated insulation with a low insulation resistance might not break down under a high potential. Since insulating members composed of different materials or combinations of materials may have inherently different insulation resistances, the numerical value of measured insulation resistance cannot properly be taken as a direct measure of the degree of cleanliness or absence of deterioration. The test is especially helpful in determining the extent to which insulating properties are affected by deteriorative influences, such as heat, moisture, dirt, oxidation, or loss of volatile materials.

Factors affecting use

Factors affecting insulation-resistance measurements include temperature, humidity, residual charges, charging currents of time constants of instrument and measured circuit, test voltage, previous conditioning, and duration of uninterrupted test voltage application (electrification time). In connection with this last-named factor, it is characteristic of certain components (for example, capacitors and cables) for the current to usually fall from an instantaneous high value to a steady lower value at a rate of decay which depends on such factors as test voltage, temperature, insulating materials, capacitance, and external circuit resistance. Consequently, the measured insulation resistance will increase for an appreciable time as test voltage is applied uninterruptedly. Because of this phenomenon it may take many minutes to approach maximum insulation-resistance readings, but specifications usually require that readings be made after a specified time, such as 1 or 2 min. This shortens the testing time considerably while still permitting significant test results, provided the insulation resistance is reasonably close to steady-state value, the current versus time curve is known, or suitable correction factors are applied to these measurements. For certain components, a steady instrument reading may be obtained in a matter of seconds. When insulation-resistance measurements are made before and after a test, both measurements should be made under the same conditions.

Apparatus

Insulation-resistance measurements should be made on an apparatus suitable for the characteristics of the component to be measured such as a megohm-meter, insulation-resistance test set, or other suitable apparatus.

Unless otherwise specified, the direct potential applied to the specimen will be that indicated by one of the following test-condition letters as specified:

Test condition	Test potential
A	100 V ± 10%
B	500 V ± 10%
C	1000 V ± 10%

For in-plant quality conformance testing, any voltage may be used provided it is equal to or greater than the minimum potential allowed by the applicable test condition. Unless otherwise stated, the measurement at the insulation-resistance value required should not exceed 10%. Proper guarding techniques are to be used to prevent erroneous readings due to leakage along undesired paths.

Procedure

When special preparations or conditions such as special test fixtures, re-connections, grounding, isolation, low atmospheric pressure, humidity, or immersion in water are required, they must be specified. Insulation-resistance measurements to be made between the mutually insulated points or between insulated points and ground, as specified. When electrification time is a factor, the insulation-resistance measurements will be made immediately after a 2 min period of uninterrupted test voltage application, unless otherwise specified. However, if the instrument-reading indicates that an insulation resistance meets the specified limit, and is steady or increasing, the test may be terminated before the end of the specified period. When more than one measurement is specified, subsequent measurements of insulation resistance will be made using the same polarity as the initial measurements.

Details

The following details must be specified:

1. Test-condition letter, or other test potential, if specified.
2. Special preparations or conditions, if required.
3. Points of measurement.
4. Electrification time, if other than 2 min.
5. Measurement error at the insulation-resistance value required, if other than 10%.

Moisture resistance

The moisture-resistance test is performed for the purpose of evaluating, in an accelerated manner, the resistance of component parts and constituent materials to the deteriorative effects of the high-humidity and heat conditions typical of tropical environments. Most tropical degradation results directly or indirectly from absorption of moisture vapour and films by vulnerable insulating materials, and from surface wetting of metals and insulation. These phenomena produce many types of deterioration, including: corrosion of metals; physical distortion and decomposition of organic materials; leaching out and spending of constituents of materials; detrimental changes in electrical properties.

This test differs from the steady-state humidity test and derives its added effectiveness in its employment of temperature cycling, which provides alternate periods of condensation and drying essential to the development of the corrosion processes and, in addition, produces a 'breathing' action of moisture into partially sealed containers. Increased effectiveness is also obtained by use of a higher temperature, which intensifies the effects of humidity. The test includes low-temperature and vibration subcycles that act as accelerants to reveal otherwise undiscernible evidences of deterioration since stresses caused by freezing mositure and accentuated by vibration tend to widen cracks and fissures. As a result, the deterioration can be detected by the measurement of electrical characteristics (including such tests as dielectric-withstanding voltage and insulation resistance) or by performance of a test for sealing.

Provision is made for the application of a polarising voltage across the insulation in order to investigate the possibility of electrolysis, which can promote eventual dielectric breakdown. The test also provides for electrical loading of certain components, if desired, in order to determine the resistance of current-carrying components, especially fine wires and contacts, to electrochemical corrosion. Results obtained with this test are reproducible and have been confirmed by investigations of field failures and have also proved reliable for indicating those parts which are unsuited for tropical field use.

Procedure

Initial conditioning
The device should be subjected to a lead-fatigue test as follows. A weight of 225 ± 15 g should be applied to each lead without restriction for one 90 degree arc of the case. An arc is defined as a movement without torsion to a position perpendicular to pull axis and return to normal.

Specimens should be tested in accordance with step tests as in Figs 6 and 7. If a specimen fails to meet requirements when tested in accordance with Fig. 7, a similar specimen should be tested in accordance with Fig. 6.

Mounting
Specimens to be mounted by their normal mounting means, in their normal mounting position.

Initial measurements
Prior to step 1 of the first cycle, the specified initial measurement to be made in ambient conditions.

Number of cycles
Specimens should be subjected to 10 continuous cycles, each as shown in Fig. 6 or Fig. 7.

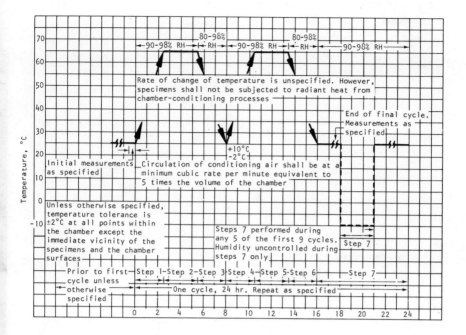

Figure 6

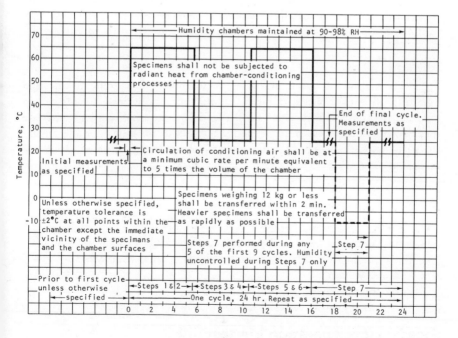

Figure 7

Subcycle

During step 7, at least 1 hr but not more than 4 hr after step 7 begins, the specimens will be either removed from the humidity chamber, or the temperature of the chamber will be reduced, for performance of step 7a. After this step, the specimens will be returned to 25°C at 90 to 98% relative humidity and kept there until the next cycle begins. This subcycle can be performed during any five of the first nine cycles.

Step 7a: at least 1 hr but not more than 4 hr after the beginning of step 7, the specimens are to be either removed from the humidity chamber, or the temperature of the chamber reduced. Specimens will then be conditioned at $-10 \pm 2°C$, with humidity not controlled, for 3 hr as indicated on Fig. 6 or Fig. 7. When a separate cold chamber is not used, care should be taken to ensure that the specimens are held at $-10 \pm 2°C$, for the full 3 hr period.

Final measurements

At high humidity

Upon completion of step 6 of the final cycle, when measurements at high humidity are specified, the specimens shall be maintained at a temperature of $25 \pm 2°C$, and an RH of 90 to 98% for a period of $1\frac{1}{2}$ to $3\frac{1}{2}$ hr, after which the specified measurements shall be made.

Due to the difficulty in making measurements under high-humidity conditions, the individual specification should specify the particular precautions to be followed in making measurements under such conditions.

After drying period

Following step 6 of the final cycle, or following measurements at high humidity if applicable, specimens should be conditioned for 24 hr in the ambient conditions specified for the initial measurements after which the specified measurements are to be made. Measurements may be made during the 24 hr conditioning period; however, failures should be based on the 24 hr period only.

Details

The following details must be specified in the individual specifications:

1 Initial measurements, and conditions if other than room ambient.
2 When applicable, the polarisation voltage if other than 100 V.
3 Loading voltage.
4 Final measurements.

Steady-state operation life test

This test is to determine compliance with the specified lambda (λ) for devices.

Procedure

The semiconductor device should be subjected to a steady-state operational-life test at a specified temperature, for 1,000 hr. The device should be operated under specified conditions. Lead mounted devices should be mounted by the leads with jig mounting clips at least 10 mm from the body or lead tubulation, if the tubulation projects from the body. The point of connection should be maintained at a temperature not less than that specified. The measurements listed under end-points for life tests should be carried out at 0 and 1,000 hr. Additional readings may be taken at the discretion of the manufacturer or user.

Details

The following details should be specified:

1 Test condition.
2 Test temperature.
3 Test mounting.
4 End-point measurements.

High-temperature life (non-operating) test

The purpose of this test is to determine compliance with the specified lambda (λ) for devices.

Procedure

The device should be stored in ambient conditions (normally maximum temperature) for 1,000 hr. At 0 and 1,000 hr the samples are to be removed from the specified ambient conditions and allowed to reach standard test conditions prior to performance of the specified end-point measurements. Additional readings may be taken at the discretion of the user.

Visual examination

The markings on the device must be legible after test and there shall be no evidence (when examined with no magnification) of flaking or pitting of the finish or corrosion that will interfere with the mechanical and electrical application of the device.

Intermittent operation life test

This test is to determine compliance with the specified lambda (λ) for devices.

Procedure

The device should be subjected, intermittently, to the specified operating and non-operating conditions for 1,000 hr total time. Lead mounted devices should be mounted by the leads with jig mounting clips at least 10 mm from the body or lead tubulation, if the lead tubulation projects from the body. The point of connection should be maintained at a temperature not less than specified. The operating interruptions for the purpose of performing the specified end-point measurements should be at 0 and 1,000 hr. The on and off periods should be initiated by sudden, not gradual, application or removal of the specified conditions.

Details

The following details should be specified:

1. Test conditions.
2. Operating and non-operating cycles.
3. End-point measurements.
4. Test temperature (case or ambient).
5. Test mounting, if other than specified.

Salt atmosphere (corrosion) test

This test is proposed as an accelerated laboratory corrosion test simulating the effects of sea-coast atmospheres on devices.

Apparatus

The apparatus used in the salt-atmosphere test should include the following:

1. Exposure chamber with racks for supporting devices.
2. Salt-solution reservoir.
3. Means of atomising the salt solution, including suitable nozzles and compressed-air supply.
4. Chamber-heating means and control.
5. Means for humidifying the air at a temperature above the chamber temperature.

Procedure

The device should be placed within a test chamber. A salt atmosphere fog having a temperature of 35°C should be passed through the chamber for a period of 24 hr. The fog concentration and velocity should be so adjusted that the rate of salt deposit in the test area is between 10,000 and 50,000 mg/m².day.

Examination

On completion of the test and to aid in examination, devices should be prepared in the following manner. Salt deposits should be removed by a gentle wash or dip in running water not warmer than 37°C and a light brushing, using a soft-hair brush or plastic bristle brush.

A device with illegible markings, evidence of flaking or pitting of the finish or corrosion that will interfere with the application of the device should be rejected.

Details

The following should be specified:
1. Time of exposure, if other than specified.
2. Measurements and examinations after test.

Salt spray (corrosion)

The salt-spray test, in which specimens are subjected to a fine mist of salt solution, has several more or less useful purposes when utilised with full recognition of its deficiencies and limitations. Originally proposed as an accelerated laboratory corrosion test simulating the effects of seacoast atmospheres on metals, with or without protective coatings, this test has been erroneously considered by many as an all-purpose accelerated corrosion test, which if 'withstood successfully' will guarantee that metals or protective coatings will prove satisfactory under any corrosive condition. Experience has since shown that there is seldom a direct reletionship between resistance to salt-spray corrosion in other media, even in so-called 'marine' atmospheres and ocean water. However, some idea of the relative service life and behaviour of different samples of the same (or closely related) metals or of protective coating-base metal combinations in marine and exposed seacoast locations can be gained by means of the salt-spray test, provided accumulated data from correlated field service tests and laboratory salt-spray tests show that such a relationship does exist, as in the case of aluminium alloys. (Such correlation tests are also necessary to show the degree of acceleration, if any, produced by the laboratory test.) The salt-spray test is generally considered unreliable for comparing the general corrosion resistance of different kinds of metals or coating-metal combinations, or for predicting their comparative service life. This test has received its widest acceptance as a test for evaluating the uniformity (specifically, thickness and degree of porosity) of protective coatings, metallic and non-metallic, and has served this purpose with varying amounts of success. In this connection, the test is useful for evaluating different lots of the same product, once some standard level of performance has been established. It is especially helpful as a screening test for revealing particularly inferior coatings. When used to check the porosity of metallic coatings, the test is more dependable when applied to coatings which are cathodic rather than anodic toward the basic metal. This test can also be used to detect the presence of free iron contaminating the surface of another metal, by inspection of the corrosion products.

Apparatus

Apparatus used in the salt-spray test includes the following:
1. Exposure chamber with racks for supporting specimens.
2. Salt-solution reservoir.

3 Means for atomising the salt solution, including suitable nozzles and compressed-air supply.
4 Chamber-heating means and control.
5 Means for humidifying the air at a temperature above the chamber temperature.

Chamber

The chamber and all accessories must be constructed of material that will not affect the corrosiveness of the fog, such as glass, hard rubber, plastic, or wood other than plywood. In addition, all parts which come in contact with test specimens are to be of materials that will not cause electrolytic corrosion. The chamber and accessories should be so constructed and arranged that there is no direct impingement of the spray or dripping of the condensate on the specimens, so that the spray circulates freely about all specimens to the same degree, and so that no liquid which has come in contact with the test specimens returns to the salt-solution reservoir. The chamber must be properly vented.

Atomisers

The atomiser or atomisers used should be of such design and construction as to produce a finely divided, wet, dense fog.

Air supply

The compressed air entering the atomisers must be free from all impurities such as oil and dirt. Means are to be provided to humidify and warm the compressed air as required to meet the operating conditions. The air pressure must be sufficient to produce a finely divided dense fog with the atomiser or atomisers used. To ensure against clogging the atomisers by salt deposition, the air should have a relative humidity of at least 85% for a 20% solution and between 95 and 98% for the 5% solution at the point of release from the nozzle. A satisfactory method is to pass the air in very fine bubbles through a tower containing heated water; the temperature of the water should be 35 °C and often higher. The permissible temperature increases with increasing volume of air and with decreasing heat insulation of the chamber and temperature of its surroundings. It should not exceed a value above which an excess of moisture is introduced into the chamber (for example, 43·3 °C at an air pressure of 12 psi), or a value which makes it impossible to meet the requirement for operating temperature.

Salt solution

The salt-solution concentration should be 5 or 20% as specified in the referencing specification. When no solution concentration is specified, the 5% solution should be used, except when this test method is referenced in issues of specifications dated prior to the approval date of this revision;

then the 20% salt solution is used. The salt used is sodium chloride containing on the dry basis not more than 0·1% of sodium iodide, and not more than 0·3% of total impurities. The 5% solution is prepared by dissolving 5±1 parts by weight of salt in 95 parts by weight of distilled or other water. The 20% solution is prepared by dissolving 20±2 parts by weight of salt in 80 parts by weight of distilled or other water. Distilled or other water used in the preparation of solutions should contain not more than 200 ppm of total solids. The solution is to be kept free from solids by filtration or decantation. The solution must be adjusted to and maintained at a specific gravity of 1·0268 to 1·0413 for the 5% solution and 1·126 to 1·157 for the 20% solution. The pH value is to be maintained between 6·5 and 7·2 when measured at a temperature between 33·9 and 36·1°C. Only 'Electronic Grade' hydrochloric acid or sodium hydroxide is to be used to adjust the pH. The pH measurement is to be made electrometrically using a glass electrode with a saturated potassium chloride bridge or by a colorimetric method such as bromothymol blue, provided the results are equivalent to those obtained with the electrometric method.

Preparation of specimens

Specimens are to be given a minimum of handling, particularly on the significant surfaces, and be prepared for test immediately before exposure. Unless otherwise specified, uncoated metallic or metallic-coated specimens should be thoroughly cleansed of oil, dirt, and grease as necessary until the surface is free from water break. The cleaning methods must not include the use of corrosive solvents or solvents which deposit either corrosive or protective films, or the use of abrasives other than a paste of pure magnesium oxide. Specimens having an organic coating should not be solvent cleaned. Those portions of specimens which come in contact with the support and, unless otherwise specified in the case of coated specimens or samples, cut edges and surfaces not required to be coated, must be protected with a suitable coating of wax or similar substance impervious to moisture.

Procedure

Location of specimens
Unless otherwise specified, flat specimens and, where practicable, other specimens, should be supported in such a position that the significant surface is approximately 15° from the vertical and parallel to the principal direction of horizontal flow of the fog through the chamber. Other specimens to be positioned so as to ensure most uniform exposure. Whenever practicable, the specimens should be supported from the bottom or from the side. When specimens are suspended from the top, suspension should be by means of glass or plastic hooks or wax string; if plastic hooks are used, they should be fabricated of material which is non-reactive to the salt solution such as lucite. The use of metal hooks is not permitted. Specimens should be positioned so that they do not contact each other, so that they do not shield each other

from the freely settling fog, and so that corrosion products and condensate from one specimen do not fall upon another.

Operating conditions

Temperature

The test shall be conducted with a temperature in the exposure zone maintained at 35°C. Satisfactory methods for controlling the temperature accurately are by housing the apparatus in a properly controlled constant-temperature room, by thoroughly insulating the apparatus and pre-heating the air to the proper temperature prior to atomisation, and by jacketing the apparatus and controlling the temperature of the water or of the air used. The use of immersion heaters for the purpose of maintaining the temperature within the chamber is prohibited.

Atomisation

The conditions maintained in all parts of the exposure zone shall be such that a suitable receptacle placed at any point in the exposure zone will collect from 0·5 to 3·0 ml/hr of solution for each 80 cm^2 of horizontal collecting area (10 cm diameter) based on an average run of at least 16 hr. When the 20% solution is used, the solution thus collected shall have a sodium chloride content of 18 to 22% (specific gravity of 1·126 to 1·157 when measured between 33·9 and 36·1°C). When the 5% solution is used, the solution thus collected shall have a sodium chloride content of 4 to 6% (specific gravity of 1·0268 to 1·0413 when measured at a temperature between 33·9 and 36·1°C). At least two clean fog-collecting receptacles should be used, one placed near the nozzle and one placed as far as possible from all nozzles. Receptacles are to be fastened so that they are not shielded by specimens and so that no drops of solution from specimens or other sources will be collected. With nozzles made of material non-reactive to the salt solution, suitable atomisation has been obtained in boxes having a volume of less than 0·34 m^3 with the following conditions:

1. Nozzle pressure of from 12 to 18 lb/in^2.
2. Orifices of from 0·02 to 0·03 inch in diameter.
3. Atomisation of approximately 3·4 litres of the salt solution per 0·28 m^3 of box volume per 24 hr.

When using large-size boxes having a volume considerably in excess of 0·34 m^3 the above conditions may have to be modified in order to meet the requirements for operating conditions.

Length of test

The length of the salt-spray test shall be that indicated in one of the following test conditions, as specified:

Test condition	Length of test, hr
A	96
B	48

Unless otherwise specified, the test shall be run continuously for the time indicated or until definite indication of failure is observed, with no interruption except for adjustment of the apparatus and inspection of the specimen.

Measurements
At the completion of the exposed period, measurements should be made as specified. To aid in examination, specimens are to be prepared in the following manner unless otherwise specified. Salt deposits are to be removed by a gentle wash or dip in running water not warmer than 37·8°C and lightly brushed, using a soft-hair brush or plastic bristle brush. The device is allowed to dry for approximately 24 hr at 40°C, prior to examination.

A device with illegible markings, evidence of flaking or pitting of finish or corrosion is to be rejected.

Details
The following details must be specified in the individual specifications:

1 Applicable salt solution.
2 Special mounting and details if applicable.
3 Test-condition letter.
4 Measurements after exposure.

Thermal shock (temperature cycling)

This test is conducted for the purpose of determining the resistance of a part to exposures at extremes of temperature, and to the shock of alternate exposures to these extremes, such as would be experienced when equipment or parts are transferred to and from heated shelters in arctic areas. These conditions may also be encountered in equipment operated non-continuously in low-temperature areas or during transportation. Although it is preferred that the specimen reach thermal stability at the temperature of a test chamber during the exposure specified, in the interest of saving test time, parts may be tested at the minimum exposure durations specified, which will not ensure thermal stability, but only approach thereto. Permanent changes in operating characteristics and physical damage produced during thermal shock result principally from variations in dimensions and other physical properties. Effects of thermal shock include cracking and delamination of finishes, cracking and crazing of embedding and encapsulating compounds, opening of thermal seals and case seams, leakage of filling materials, and changes in electrical characteristics due to mechanical displacement or rupture of conductors or of insulating materials.

Apparatus

Separate chambers should be used for the extreme temperature conditions of steps 1 and 3. The air temperature of the two chambers is to be held at each of the extreme temperatures by means of circulation and sufficient hot or cold chamber thermal capacity so that the ambient temperature should reach the specified temperature within 2 min after the specimens have been transferred to the appropriate chamber.

Procedure

Specimens are to be placed in such a position with respect to the airsteam that there is substantially no obstruction to the flow of air across and around the specimen. When special mounting is required, it should be specified. The specimen is then to be subjected to the specified test condition of Table 1, for a total of five cycles performed continuously. One cycle consists of steps 1 to 4 of the applicable test condition. Specimens are not to be subjected to forced circulating air while being transferred from one chamber to another. Direct heat conduction to the specimen should be minimised.

Measurements

Specified measurements are made prior to the first cycle and upon completion of the final cycle, except that failures are to be based on measurements made after the specimen has returned to thermal stability at room ambient temperature following the final cycle.

Details

The following details must be specified in the individual specification:

1. Special mounting, if applicable.
2. Test-condition letter.
3. Measurements before and after cycling.

Table 3 Thermal shock test conditions

Step	Test condition A Temperature, °C	Time, min	Test condition B Temperature, °C	Time, min	Test condition C Temperature, °C	Time, min
1	−55(+0, −3)	(See Table 4)	−65(+0, −5)	(See Table 4)	−65(+0, −5	(See Table 4)
2	25(+10, −5)	5 max.	25(+10, −5)	5 max.	25(+10, −5)	5 max.
3	85(+3, −0)	(See Table 4)	125(+3, −0)	(See Table 4)	200(+5, −0)	(See Table 4)
4	25(+10, −5)	5 max.	25(+10, −5)	5 max.	25(+10, −5)	5 max.
1	−65(+0, −5)	(See Table 4)	−65(+0, −5)	(See Table 4)	−65(+0, −5)	(See Table 4)
2	25(+10, −5)	5 max.	25(+10, −5)	5 max.	25(+10, −5)	5 max.
3	350(+5, −0)	(See Table 4)	500(+5, −0)	(See Table 4)	150(+3, −0)	(See Table 4)
4	25(+10, −0)	5 max.	25(+10, −5)	5 max.	25(+10, −5)	5 max.

Table 4 Exposure time at temperature extremes

Weight of specimen, kg	Minimum time (for steps 1 and 3), hr
0.13 and below	0.5
0.13 to 1.3	1
1.3 to 13	2
13 to 130	4
About 130	8

Thermal shock (glass strain) test

This test is to determine the resistance of the device to sudden exposure to extreme changes in temperatures.

Procedure

The device should be subjected to one of the following test conditions. There must be no eivdence of mechanical damage at the conclusion of this test.

Test condition 'A'

The device should be preconditioned by being immersed and be in intimate contact with a suitable liquid at a temperature of 100°C for a minimum of 15 sec. Immediately on conclusion of the preconditioning time, the device is to be transferred to a liquid at a temperature of 0°C and the device to remain at a low temperature for a minimum of 5 sec and then transferred to a liquid at a temperature 100°C and to remain at the high temperature for a minimum of 15 sec.

Transfer time from high temperature to low temperature and from low to high temperature should be less than 3 sec. Unless otherwise specified, the duration of the test should be 5 complete cycles as illustrated in Fig. 8.

Test condition 'B'

Test condition 'B' should be the same as test condition 'A', except that the time at the high temperature should be a minimum of 5 min, the time at the low temperature a minimum of 5 min, and the transfer time should be less than 10 sec.

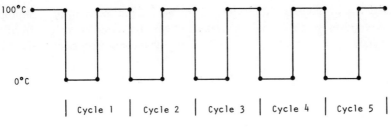

Figure 8

Details

The following details should be specified:

1 Test condition letter.
2 Number of cycles if other than 5.

Temperature measurement: case and stud

This method measures case temperatures of hex-base devices.

Type of thermocouple

The thermocouple material should be copper-constantan, for the range of 190 to 350°C. The junction of the thermocouple should be welded together to form a bead rather than soldered or twisted.

Accuracy

An accuracy of 0·5 degC should be expected of the thermocouple. Under load conditions slight variations in the temperature of different points on the case may reduce the accuracy to $\pm 1\cdot 0$ degC for convection cooling and $\pm 2\cdot 0$ degC for forced air ventilation.

Procedure

Method of mounting
A small hole, just large enough to insert the thermocouple should be drilled approximately 1 mm deep into the flat of the case hex at a point chosen. The edge of the hole should then be peened with a small centre punch to force a rigid mechanical contact with the welded bead of the thermocouple. If forced air ventilation is used, the thermocouple should be mounted away from the air stream and the thermocouple leads close to the junction shielded. Other methods of mounting the thermocouple, with the possible exception of the thermocouple welded directly to the case, will result in temperature readings lower than the actual temperature. These deviations will result from:

1 Inadequate contact with case using cemented thermocouples.
2 External heat sink in contact with the thermocouple using pressure contacts.

Dew point test

This test is to monitor the device parameter for a discontinuity. The apparatus used in this test should be capable of varying the temperature from the specified high temperature to −65°C while the parameter is being measured.

Procedure

The voltage and current specified for the device should be applied to the terminals and the parameter monitored from a specific high temperature to −65°C. The dew point is indicated by a sharp discontinuity in the parameter being measured with respect to temperature. If no discontinuity is observed, it should be assumed that the dew point is at a temperature lower than −65°C and the device being tested is acceptable.

Details

The following details should be specified:

1 Test temperature.
2 Test voltage and current.
3 Test parameter.

Chapter 3 Mechanical tests

Constant acceleration test

The constant acceleration test is used to determine the effects on devices of a centrifugal force. This test is an accelerated test designed to indicate types of structural and mechanical weaknesses not necessarily detected in shock and vibration tests. Constant acceleration tests should be carried out with apparatus capable of meeting the minimum requirements of individual specifications.

Procedure

The device should be restrained by its case, or by normal mountings and the leads or cables secured. A centrifugal acceleration of the value specified should then be applied to the device for 1 min in each of the orientations X_1, X_2, Y_1, Y_2, Z_1 and Z_2. The acceleration should be increased gradually, to the valve specified, in not less than 20 sec. The acceleration should be decreased gradually to zero in not less than 20 sec.

Details

The following details should be given:

1 Amount of centrifugal force to be applied, in gravity units (g).
2 Measurements to be made after test.

Shock test

The shock test is intended to determine the suitability of the device, with or without auxiliary protection, for use in electronic equipment which may be subjected to moderately severe shocks as a result of suddenly applied forces or abrupt changes in motion produced by rough handling, transportation or field service or operation. Shocks of this type may disturb operating characteristics or cause damage similar to that resulting from excessive vibration, particularly if the shock pulses are repetitive.

The shock-testing equipment should be capable of providing minimum shock pulses of $1500g$ peak with a pulse duration of approximately 0·5 msec, to the body of the device.

Procedure

The shock-testing apparatus should be mounted on a sturdy laboratory table or equivalent base and levelled before use. The device should be rigidly mounted or restrained by its case with a suitable protection for the leads. For each blow, the carriage is to be raised to the height necessary for obtaining the specified acceleration and then allowed to fall. Means may be provided to prevent the carriage from striking the anvil a second time.

Details

The following details should be specified:

1. Acceleration and duration of pulse.
2. Number and direction of blows.
3. Electrical-load conditions, if applicable.
4. Measurements after shock.
5. Measurements during shock.

Shock: simulated drop test

This test is intended to determine the suitability of the devices to withstand a severe shock, by providing for the application of such shocks directly to the devices.

Procedure

The device must be rigidly mounted or restrained by its case with suitable protection for the leads. The device should then be subjected to the specified number of blows, each with an acceleration of 3000g peak and a pulse of approximately 0·2 msec, in each of the specified directions.

Details

The following details should be specified:

1. Number and direction of blows.
2. Measurement after test.
3. Measurement during drop.

Solderability

The reason for this test method is to determine the solderability of all solid and stranded wires up to 4 mm thickness, and lugs, tabs, hook leads, turrets, etc., which are normally joined by a soldering operation. This determination is made on the basis of the ability of these terminations to be wetted by a new coating of solder, or to form a suitable fillet when dip soldered to a specially prepared solderable wire.

The procedures will verify that the treatment used in the manufacturing process to facilitate soldering is satisfactory and that it has been applied to the required portion of the part which is designed to accommodate a solder connection. An accelerated aging is included in this test method which simulates a minimum of 6 month's natural aging under a combination of various storage conditions that have different deleterious effects.

Apparatus

Solder pot
A solder pot of sufficient size to contain at least 1 kg of solder and capable of maintaining the solder at the temperature specified.

Dipping device
A mechanical dipping device capable of controlling the rates of immersion and emersion of the terminations and providing a dwell time (time of total immersion to the required depth) in the solder bath.

Optical equipment
An optical system having a magnification of ten diameters to be used.

Container and cover
A non-metallic container of sufficient size to allow the suspension of the specimens 4 cm above the boiling distilled water. (A 2,000 ml beaker is one size that has been used satisfactorily for smaller components.) The cover to be of one or more stainless steel plates and capable of covering approximately seven-eighths of the open area of the container so that a more constant temperature may be obtained.

As a suitable method of suspending the specimens, perforations or slots in the plates are permitted for this purpose.

Materials

Flux
A flux, soldering, liquid (rosin base).

Solder
Tin alloy; lead-tin alloy; lead alloy.

Standard copper wrapping wire
The standard wrapping wire to be fabricated from type S, soft or drawn and annealed, uncoated. The diameter of the wrapping wire to be 0·635 mm. The preparation of the wrapping wire will be as follows:

1 Straighten and cut wire into convenient lengths (5 cm minimum).
2 Degrease by immersion in either acetone or trichloroethane for 2 min.
3 Clean in HBF (10% water solution) for 5 min with agitation. Use caution in handling.
4 Rinse acid off as follows: (a) two cold water rinses; (b) two isopropyl alcohol rinses.
5 Immerse in flux.
6 Dip in molten solder for 5 sec at $230\pm5°C$ ($450\pm10°F$).
7 Rinse in isopropyl alcohol.

8 Standard wrapping wire to be sorted in clean, covered container if not used immediately. If not used within 48 hr after preparation it should be reprocessed as specified.
9 The usable life of the standard wrapping wire must not exceed 48 hr.

Procedure

The test procedure is to be performed on the number of terminations. During handling care is to be exercised to prevent the surface to be tested from being abraided or contaminated by grease, perspirants, etc. The test procedure will consist of the following operations:
1 Proper preparation of the specimen if applicable.
2 Aging of all specimens.
3 Application of standard wrapping wire where applicable.
4 Application of flux and immersion of the terminations into molten solder.
5 Examination and evaluation of the tested portions of the terminations upon completion of the solder-dip process.

Preparation of terminations
No wiping, cleaning, scraping, or abrasive cleaning of the terminations shall be performed. Any special preparation of the terminations, such as bending or reorientation prior to the test shall be specified in the individual specification. If the insulation on stranded wires must be removed, the method should not loosen the strands in the wire.

Aging
Prior to the application of the flux and subsequent solder dips, all specimens assigned to this test are to be subjected to aging by exposure of the surfaces to be tested to steam in the container. The specimens to be suspended so that no portion of the specimen is less than 4 cm above the boiling distilled water for 60 min. Means of suspension shall be a non-metallic holder. If necessary, additional hot distilled water may be gradually added in small quantities so that the water will continue to boil and the temperature will remain essentially constant.

Application of standard solderable wire for lugs, tabs
All aged specimens to have a wrap of $1\frac{1}{2}$ turns of the standard wire around the portion of the specimen to be tested. The standard wrapping wire to be wrapped in such a manner so that it will not move during the solder dip.

Application of flux
Terminations to be immersed in the flux, which is at room temperature, to the minimum depth necessary to cover the surface to be tested. Unless otherwise specified in the individual specification, terminations will be immersed to within 0·125 cm of the body of the part, for a period of from 5 to 10 sec.

Solder dip

The dross and burned flux is to be skimmed from the surface of the molten solder.

The molten solder to be stirred with a clean, stainless-steel paddle to assure that it is at a uniform temperature of 230±5°C. The surface of the molten solder to be skimmed again just prior to immersing the terminations in the solder. The part will be attached to a dipping device and the flux-covered terminations immersed once in the molten solder to the same depth. The immersion and immersion rates to be 2·5+0·6 cm/sec and the dwell time in the solder bath of 5·0±0·5 sec. After the dipping process, the part to be allowed to cool in air. The residue flux to be removed from the terminations by dipping in clean isopropyl alcohol. If necessary, a soft damp cloth moistened with clean 91% isopropyl alcohol can be used to remove all remaining flux.

Examination of terminations

After each dip-coated termination has been thoroughly cleaned of flux, the 2·5 cm portion of the dipped lead nearest the component, or the whole lead if less than 2·5 cm, or the fillet area (whichever is applicable), is to be examined using a magnification of 10 diameters.

Evaluation of solid wire terminations

The criteria for acceptable solderability during the evaluation of the terminations are:

1. The termination is at least 95% covered by a continuous new solder coating.
2. Pinholes or voids are not concentrated in one area and do not exceed 5% of the total area.

Evaluation of lugs, tabs

The criteria for acceptable solderability during evaluation of the terminations and wires are:

1. That 95% of the total length of fillet, which is between the standard wrap wire and the termination, be tangent to the surface of the termination being tested and be free from pinholes, voids, etc.
2. That a ragged or interrupted line at the point of tangency between the fillet and the termination under test is considered a defect.

The following details must be specified in the individual specification:

1. The number of terminations of each part to be tested.
2. Special preparation of the terminations, if applicable.
3. Depth of immersion if other than 0·125 cm.
4. Solder dip.
5. Examination of terminations.
6. Measurements after test.

Soldering heat

To determine a device's resistance to the high temperature encountered during soldering it is necessary to employ a temperature controlled solder pot. The following procedure should be followed. The leads of the device are immersed for 10 sec in molten metal, without flux, at a temperature of 260°C, to a point 1·6 mm from the body, tubulation, or stud of the device. One immersion for each lead of the device constitutes one cycle. The number of cycles is governed by the specific requirement of the design.

All leads may be immersed simultaneously at the discretion of the user, but the device should be allowed to return to ambient temperature between cycles. The test results to include the number of cycles also the measurements after test.

Terminal strength test

Test condition No. 1: tension

This test is designed to check the capabilities of the device leads, welds, and seals to withstand a straight pull. The tension test requires suitable clamps, vice and hand vice for securing the device and for securing the specified weight to the device lead without lead restriction.

Procedure

The specified weight should be applied without shock, to each lead or terminal. The case of the device should be held in a fixed position. When testing axial lead devices, the device is to be supported, with the leads in a vertical position, by securing one lead to a clamp or vice. With a hand vice or equivalent, the specified weight, including the attaching device should be fastened to the lower lead for the time specified. Each lead should be fastened as close to its end as practicable. When examined after removal of the stress using $\times 10$ magnification, loosening or relative motion between the terminal lead and the device body is considered a device failure.

Details

The following details should be specified:

1. Weight to be attached to lead.
2. Length of time weight is to be attached.
3. Measurements to be made after this test.

Test condition No. 2: lead or terminal torque

This test checks device leads and seals for their resistance to twisting motions. The torque test requires suitable clamps and fixtures and a torsion wrench or other suitable method of applying the specified torque without lead restriction.

Procedure
With the body of the device securely clamped, with a suitable fixture and the specified torque applied to the portion of the terminal nearest the seal for a specified time. The specified torque should be applied, without shock, about the device axis. The torque should be applied between the lead or terminal and the case, in a direction which tends to cause loosening of the lead or terminal.

UHF and microwave diodes
Unless otherwise stated, a torque of 1·7 kg cm about the diode axis should be applied for a specified time, without shock, between the terminals, and in a direction which tends to cause loosening of the terminals.

The method of clamping should be agreed with the manufacturer of the device. Examine the device using $\times 10$ magnification after removal of stress, any evidence of breakage (other than meniscus), loosening, or relative motion between the terminal lead and the device body should be considered a device failure.

Details
The following details are to be specified:

1 The amount of torque to be applied.
2 Length of time torque.
3 Measurements to be made after test.

Stud torque tests

This test is designed to check the resistance of the device with threaded mounting stud to the stress caused by tightening the device when mounting. The torque test requires suitable clamps and fixtures and a torsion wrench or a suitable method of applying the torque.

Procedure

The device should be clamped by its body or flange. A flat washer of 1·6 mm minimum thickness and a new class 2 fit nut should be assembled in that order on the stud, with all parts clean and dry. A torque should be applied for the specified length of time without shock to the nut. The nut and washer should be disassembled from the device and examined for compliance of requirements. The device is considered a failure if:

1 The stud breaks or the device exhibits any visual elongation or mechanical deformation.
2 If it fails post-test end-point measurements.

Details
The following details should be specified:

1 The amount of torque to be applied.

2 Length of time the torque to be applied.
3 Measurements to be made after test.

Lead fatigue test

This test is to check the resistance of the device to metal fatigue. The lead-fatigue test is made using a weight and with a suitable clamping or attaching devices.

Procedure

Where applicable, two leads on each device are to be tested. The leads selected in a cyclical manner (regular recurring), when applicable, that is, leads number 1 and 2 on the first device, number 2 and 3 on the second device. A weight of 227 g should be applied to each lead for three 90° arcs of the case. An arc is defined as the movement of the case, without torsion, to a position perpendicular to the pull axis and return to normal. All arcs on a single lead will be made in the same direction and in the same plane without lead restriction. One bending cycle is to be completed in from 2 to 5 sec. Any glass seal fracture (other than the meniscus) or broken lead is to be considered a failure.

Details

The following details should be given:

1 Weight to be attached to the lead, if other than 227 g.
2 Number of arcs, if other than three.
3 Measurements to be made after this test.

Bending stress test

This test is made to check the quality of the leads, lead welds and glass-to-metal seals of the device. Bending stress tests are made using attaching devices, such as suitable clamps or other supports for stud-mounted devices.

Procedure

1 Cylindrical devices. With one contact of the device held in a suitable clamp, the specified force should be applied, without shock, at right angles to the reference axis of the device, as near the top of the opposite contact or tubulation as practicable.
2 Stud-mounted devices. The device should be securely fastened, with its reference axis in a horizontal position, by screwing the stud into a suitable support. With a hand vice or equivalent a weight should be suspended from the hole in the lug for a given length of time.

When examined after removal of stress using $\times 10$ magnification, any evidence of breakage, loosening, or relative motion between the terminal lead and the device body should be considered a failure.

Details

The following details are to be specified:

1. Special preparations or conditions.
2. Weight to be attached to lead.
3. Length of time weight is applied.
4. Measurements to be made after test.

Vibration, variable frequency (monitored) test

This test is performed for the purpose of determining the effect of vibration on semiconductor devices in a specified frequency range at a specific acceleration. This test may also be useful in detecting foreign particles within the device.

The device is to be rigidly fastened on the vibration platform. Special care is required to ensure positive electrical connection to the device leads to prevent intermittent contacts during vibration. Care must also be exercised to avoid magnetic fields in the area of the device being vibrated. The device should be vibrated with a simple harmonic motion with a constant peak acceleration of $20g$ minimum. The acceleration to be monitored at a point where the g level is equivalent to that of the support point for the devices. The vibration should be varied logarithmically between 50 and 2000 Hz. The entire frequency range of 50 to 2000 Hz and return to 50 Hz should be traversed in not less than 8 min. This frequency range should be executed one time in each of the orientations X, Y and Z (total 3 times) so that the motion is applied for a total of 24 min minimum.

Interruptions are permitted provided the requirements for the rate of change and test duration are met. Completion of vibration within any separate frequency band can be made before going on to the next band.

With the specified d.c. voltages and currents applied the device should be monitored continuously, during the vibration period, for intermittent opens or shorts. The monitoring device should be capable of detecting voltage or current changes of 10 μsec or more duration and of the magnitude specified. In addition, it should be a positive-indication 'go-no-go' device or a recorded trace. Devices requiring continuous visual monitoring, such as an oscilloscope, should not be used. The circuit should be sensitive enough to detect a 10 μA change.

Details

The following details are to be specified:

1. The test circuit or the voltage, currents and impedance needed to establish the test.
2. The voltage or current-change and its limits to be monitored.
3. The device to be reverse-biased at one-half of its rated voltage.
4. Post-test measurements.

Vibration fatigue test

This test is to determine the effect on the device of vibration within a given frequency range.

Procedure

The device should be rigidly fastened on a vibration platform and the leads or cables adequately secured.

The device is then to be subjected to a simple harmonic motion in the range of 60 Hz, with a constant peak acceleration of 20g minimum. The vibration should be applied for 32 hr minimum, in each of the orientations X_1, Y_1 and Z_1 for a total of 96 hr minimum.

Details

The following details to be specified:

1. Test voltage and current.
2. Test frequency.
3. Peak acceleration.
4. Test time, if required.

Vibration, variable frequency test

The variable-frequency, vibration test is performed for the purpose of determining the effect on component parts of vibration in a certain frequency range.

The device should be rigidly fastened on the vibration platform and the leads or cables adequately secured. The device is to be subjected to a constant peak acceleration of 20g minimum, the vibration frequency to be varied approximately logarithmically between 100 and 2000 Hz.

The entire frequency range of 100 to 2000 Hz and return to 100 Hz to be traversed in not less than 4 min. This cycle to be performed four times in each of the orientations X_1, Y_1 and Z_1 (total 12 times), so that the motion is applied for a total period of approximately 48 min, measurements after test to be specified.

Vibration noise test

This test measures the amount of electrical noise produced by the device under vibration.

Procedure

The device and its leads should be rigidly fastened on the vibration platform and the leads or cables adequately secured. The device is to be vibrated with simple harmonic motion with a constant peak acceleration of 20g minimum. The vibration frequency to be varied approximately logarithmically between

100 and 2000 Hz. The entire frequency range to be traversed in not less than 5 min for each cycle. This cycle should be performed once in each of the orientations X_1, Y_1 and Z_1 (total of three times), so that the motion is applied for a total period of approximately 12 min. The specified voltages and currents should be applied in the test circuit. The maximum noise-output across the specified load resistor during traverse should be measured with an average-responding root-mean-square (r.m.s.) calibrated high-impedance voltmeter. The meter should measure, with an error of not more than 3%, the r.m.s. value of a sine wave voltage at 200 Hz. The characteristic of the meter over a bandwidth of 20 to 2000 Hz should be ± 1 dB of the value at 2000 Hz, with an attenuation rate below 20 and above 2000 Hz of 6 ± 2 dB/octave. The maximum inherent noise in the circuit should be at least 10 dB below the specified noise-output voltage.

Details

The following details should be specified:

1 Test voltages and currents.
2 Load resistance.
3 Post test measurement.
4 Noise-output voltage limit.

Chapter 4 Semiconductor (transistor) electrical tests

Collector-to-base breakdown voltage

This test is carried out to determine if the breakdown voltage of the device under design conditions is greater than the specified limit.

Test circuit

In Fig. 9, R_1 is a current-limiting resistor and should be of sufficiently high resistance to avoid an excessive current flowing through the device also the current measuring meter. The voltage should be gradually increased, with

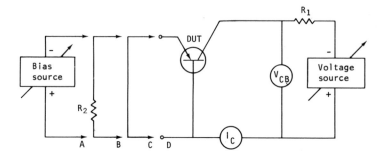

Figure 9 Test circuit. The ammeter should present essentially a short-circuit to the terminal between which the current is being measured or the voltmeter readings should be corrected for the ammeter drop

the specified bias conditions A, B, C or D applied, from zero until either the minimum limit for BV_{CBX} or a specified test current is reached. The device is acceptable if the minimum limit for BV_{CBX} is reached before the test current reaches the specified value. If the specified test current is reached first, then the device should be rejected.

Details

The following details are to be specified:
1. Test current.
2. Bias condition.
3. Condition A—Emitter to base: reverse bias.
   Condition B—Emitter to base: resistance return.
   Condition C—Emitter to base: short-circuit.
   Condition D—Emitter to base: open-circuit.

Burnout by pulsing

This test is to determine the capabilities of a device to withstand pulses. The device is to be subjected to a pulse or pulses of a length, voltages and/or currents and repetition rate specified with the design pre-pulsed conditions.

Details

The following details are to be specified:

1 Pre-pulse conditions.
2 Pulse width.
3 Pulse voltages/or currents.
4 Repetition rate.
5 Measurements after test.
6 Length of test or number of cycles.

Collector-to-emitter breakdown voltage

This test will determine if the breakdown voltage of the device is greater than the specified minimum limit.

Test circuit

In Fig. 10 the resistor R_1 is the current-limiting resistor and is to be of a sufficiently high resistance to avoid excessive current flowing through the device, also the current meter.

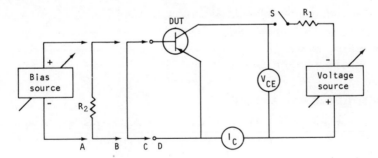

Figure 10 Test circuit. Note:
(1) If necessary switch S should be used to provide pulses of short duty cycle to minimise the rise in function temperature.
(2) The ammeter presents essentially a short-circuit to the terminals between which the current is being measured or the voltmeter readings should be corrected for the ammeter drop

Procedure

The voltage should be increased, with the specified bias conditions (conditions A, B, C, or D) applied, until the specified test current is reached. A device is acceptable if the voltage at the specified test current is greater than the minimum limit for BU_{CEX}. If desirable, a suitable base current may be applied as the voltage is increased; however, the specified bias condition and test current must be applied when the voltage is measured. This method is intended for devices exhibiting negative resistance breakdown characteristics. In such cases, extreme care must be taken to ensure that the collector current and the junction temperature of the device remain at a safe value.

Details

The following details should be specified:

1. Test current.
2. Duty cycle and pulse width, when required.
3. Bias condition as follows:

    A Emitter to base: reverse bias (specify bias voltage).
    B Emitter to base: resistance return (specify value of R_2)
    C Emitter to base: short-circuit.
    D Emitter to base: open-circuit.

Drift

This test is to determine the drift of a parameter specified in the detail of a device. The apparatus used is the same as that used for testing the associated parameters. Voltages and current as specified should be applied. In a period from 10 sec to 1 min the drift should be no more than specified in the detailed test specifications.

Details

The following details should be given:

1. Test currents and voltages.
2. Test parameter.
3. Test apparatus or test circuit.

Floating potential test

This test measures the d.c. potential between the specified, open-circuit terminal and the reference terminal when a d.c. potential is applied to other specified terminals.

Test circuit

The circuit shown in Fig. 11 is for measuring the emitter potential. For other device configurations the circuitry should be modified so that it is capable of demonstrating the device conformance to the minimum design requirements.

Procedure

The specified d.c. voltage is to be applied to the specified terminals and the d.c. voltage of the open-circuited terminal and the reference terminal monitored.

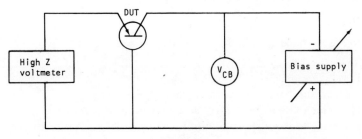

Figure 11

Details

The following details should be recorded:

1. Test voltage.
2. Input resistance of high impedance voltmeter.
3. Test, voltage application and reference terminals.

Breakdown voltage, emitter-to-base test

This test is required to determine if the breakdown voltage of the device is greater than the specified minimum limit.

Test circuit

It should be noted that the ammeters should present essentially a short-circuit to the terminals between which the current is being measured or the voltmeter readings should be corrected for the ammeter drop.

Procedure

The resistor R_1 (see Fig. 12) is a current-limiting resistor and should be of sufficiently high resistance to avoid excessive current flowing through the device and current meter. The voltage should be gradually increased, with condition (A, B, C or D) applied, from zero until either the minimum limit for BU_{EBX} or the specified test current is reached. The device is acceptable if the minimum limit for BU_{EBX} is reached before the test current reaches the specific value. If the specified test current is reached first, the device is rejected.

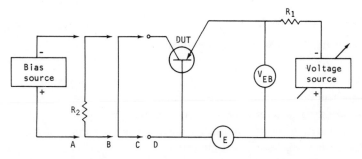

Figure 12

Details

The following details are to be specified:

1. Test current.
2. Bias condition.
3. Collector to base: reverse bias (specify bias voltage).
4. Collector to base: resistance (specify resistance of R_2).
5. Collector to base: short-circuit.
6. Collector to base: open-circuit.

Collector-to-base current test

This test measures the cut-off current of the device under the specified conditions.

Test circuit

The ammeter in this circuit (Fig. 13), must present essentially a short-circuit to the terminals between which the current is being measured or the voltmeter should be corrected for the drop across the ammeter.

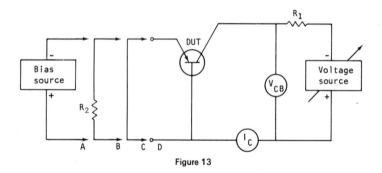

Figure 13

Procedure

The specified d.c. voltage should be applied between the collector and the base with condition A, B, C or D applied to the emitter. The measurement of current should be made at the specified ambient or case temperature.

Details

The following details should be specified.:

1. Test voltage.
2. Test temperature if other than 25°C and if case or ambient.
3. Bias condition:

   A—Emitter to base: reverse bias, state bias voltage.
   B—Emitter to base: resistance return, specify resistance of R_2.
   C—Emitter to base: short-circuit.
   D—Emitter to base: open-circuit.

Collector-to-emitter voltage test

This test measures the voltage between the collector and emitter of a device.

Procedure

The bias supplies (see Fig. 14) should be adjusted until specified voltages and currents are achieved. The voltage between the collector and emitter should be measured. If high current values are to be used in this measurement, suitable pulse techniques may be used to provide pulses of short duty cycle to minimise the rise in junction temperature.

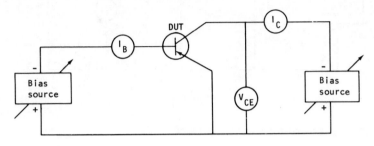

Figure 14

Details

Details to be specified are:
1 Test voltages and currents.
2 Duty cycle and pulse width if applicable.

Collector-to-emitter cut-off current test

The purpose of this test is to measure the cut-off of the device.

Test circuit

The ammeter should present substantially a short-circuit to the terminals between which the current is being measured or that the voltmeter should be corrected for the drop across the ammeter.

Procedure

The specified voltage should be applied (see Fig. 15) between the collector and emitter with a specified bias condition, A, B, C or D, applied to the base. The measurement of current should be made at the specified ambient or case temperature.

Details

Details to be specified:
1 Test voltage.
2 Test temperature if other than 25°C and whether case or ambient.

3 Bias condition:

A—Emitter to base: reverse bias (specify bias voltage).
B—Emitter to base: resistance return (specify resistance value of R_2).
C—Emitter to base: short-circuit.
D—Emitter to base: open-circuit.

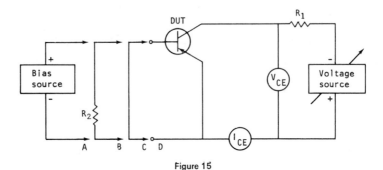

Figure 15

Collector-to-base voltage test

This test measures the voltage between collector and base under specified conditions.

Procedure

With the bias supplies (Fig. 16) adjusted until the specified voltages and currents are achieved. The voltage between the collector and base is measured. If high current values are to be used in this measurement, suitable pulse techniques may be used to provide pulses of short dury cycle to minimise the rise in junction temperature.

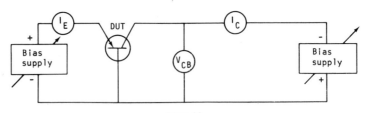

Figure 16

Details

The following details are to be specified:

1 Test voltages and currents.
2 Duty cycle and pulse width, if applicable.

Emitter-to-base current test

This test measures the cut-off current of the device under specified conditions. The ammeter in Fig. 17 should present essentially a short-circuit to the terminals between which the current is being measured, or the voltmeter corrected for the drop across the ammeter.

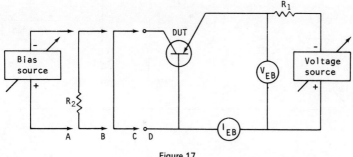

Figure 17

The specified direct current voltages should be applied between the emitter and the base with the specified conditions, A, B, C or D, applied to the collector. The measurements of current to be made at ambient or case temperature.

Details

The following details are to be specified:
1. Test voltage.
2. Test temperature if other than 25°C and whether ambient or case temperature.
3. Bias condition:
   A—Collector to base: reverse bias (specify bias voltage)
   B—Collector to base: resistance return (specify resistance R_2).
   C—Collector to base: short-circuit.
   D—Collector to base: open circuit.

Base-to-emitter voltage (saturated or non-saturated) test

This test measures the base-to-emitter voltage of the device in either a saturated or non-saturated condition.

Test circuit

The circuit (Fig. 18), and the procedure are for base to emitter. For other parameters the circuit and procedure should be changed accordingly.

Procedure

It should be noted that if necessary, switch 'S' should be used to provide pulses of short-duty cycle to minimise the rise in junction temperature. When pulsing techniques are used, oscillographic methods should be used to measure V_{BE} and other necessary parameters and the duty cycle and pulse width specified.

52 Semiconductor devices

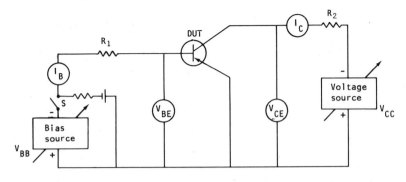

Figure 18

Test condition A (saturated)
Resistance R_1 should be made large. If the pulse method is used, R_2 should be chosen in combination with V_{CC} so that the specified collector is achieved at a value of V_{CC} low enough to ensure that the device will not be operated in breakdown between pulses. If the pulse method is not used, R_2 can be any convenient value. The current I_B and the voltage V_{CC} should be adjusted until I_B and I_C achieve their specified value. Then $V_{BE} = V_{BE(sat)}$.

Test condition 'B' (non-saturated)
For this test R_2 should be zero. The specified values of I_B and V_{CE} should be applied. V_B is then measured, alternately, the specified V_{CE} should be applied and I_B adjusted to obtain the specified I_C.

Details

The following details to be recorded:

1. Duty cycle and pulse width, when required.
2. Test condition letter.
3. Test voltages and currents.
4. Parameter to be measured.

Saturation voltage and resistance test

The purpose of this test is to measure the voltage and the resistance of the device under specified conditions.

Test circuit

The circuit (Fig. 19) and the procedure shown are for collector to emitter. For other parameters the circuit and procedure should be changed accordingly. If necessary, switch S should be used to provide pulses of short duty cycle to minimise the rise in junction temperature. When pulsing techniques are used, oscillographic methods shoud be used to measure V_{BE} and the other necessary parameters and the duty cycle and pulse width specified.

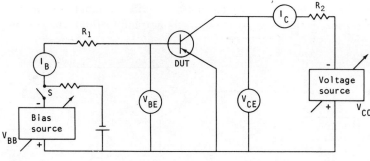

Figure 19

Procedure

Resistance R_1 should be made large. If the pulse method is used, R_2 is to be chosen in combination with V_{CC} so that the specified collector current may be achieved at a value of V_{CC} which is low enough to ensure that the device is not operated in breakdown between pulses. If pulse methods are not used, R_2 may be any convenient value. The current V_C and V_{CC} should be adjusted until I_B and I_C achieve their specified value. $V_{CE(sat)}$ is then equal to the voltage measured by voltmeter V_{CE} under specified conditions. Saturation resistance may be determined from the same circuit conditions as follows:

$$r_{CE(sat)} = V_{CE(sat)}/I_C$$

Details

The following details should be specified:

1 Duty cycle and pulse width, when required.
2 Test voltages or currents.
3 Parameter to be measured.

Forward-current transfer ratio test

This test measures the forward-current transfer ratio of the device.

Test circuit

Circuit and procedure shown (Fig. 20) are for common emitter. For other parameters the circuit and procedure should be changed accordingly. The ammeter should present essentially a short-circuit to the terminals between which the current is being measured or the voltmeter should be corrected for the drop across the ammeter.

Procedure

The voltage V_{CE} should be set to the specified value and the current I_B adjusted until the specified I_C is achieved. Then $h_{FE} = I_C/I_B$. If high current values are to be used in this measurement, switch 'S' is to be used to provide pulses of short-duty cycle to minimise the rise in junction temperature. When pulsing techniques are used, oscillographic methods may be used to measure I_C and I_B.

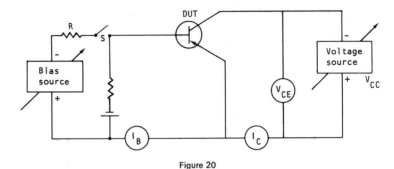

Figure 20

Details

The following details should be specified:

1. Test voltage or current.
2. Duty cycle and pulse width, when required.
3. Parameter to be measured.

Static input resistance test

The purpose of this test is to measure the input resistance of the device under specified conditions.

Test circuit

The circuit (Fig. 21) and the procedure shown are for common emitter. For other parameters, the above should be changed accordingly. If necessary, switch 'S' should be used to provide pulses of short duty cycle to minimise the rise in junction temperature. When pulsing techniques are used to measure V_{BE} and other necessary parameters, oscillographic methods should be used and the duty cycle and pulse width specified.

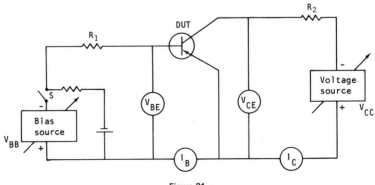

Figure 21

Procedure

Resistance R_1 should be made large. If the pulse method is used, R_2 should be chosen in combination with V_{CC} so that the specified collector current is achieved at a value of V_{CC} low enough to ensure that the device will not be operated in breakdown between pulses. If the pulse method is not used, R_2 can be any convenient value. The current I_B and V_{CC} should be adjusted until I_B and I_C achieve their specified values. Then

$$h = V_{BE}/I_B$$

Details

The following details should be specified:

1. Pulse duty cycle and width, when required.
2. Test voltages or currents.
3. Parameter to be measured.

Static transconductance test

This test measures the static transconductance of the device.

Test circuit

For other configurations, the circuit (Fig. 22) may be modified in such a manner that it is capable of demonstrating device conformance to the minimum requirements of the individual specification.

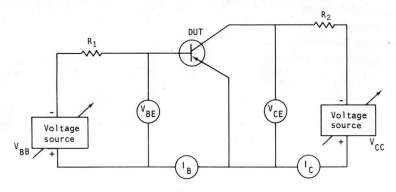

Figure 22

Procedure

Resistance R_1 should be made large or the voltage source V_{BB} replaced by a constant current source. Resistance R_2 should be chosen in combination with V_{CC} so that the specified collector current is achieved at a value of V_{CC} which is lower than BV_{CEO}. The current I_B should be adjusted until V_{CE} and I_C achieve their specified values. The current I_C or I_E and the voltages V_{BE}, V_{BC} or V_{EB} are to be measured. Using the values obtained through these measurements, the static transconductance should be calculated as follows:

For common emitter:
$$g_{ME} = I_C/V_{BE}$$

For common collector:
$$g_{MC} = I_E/V_{BC}$$

For common base:
$$g_{MB} = I_C/V_{EB}$$

If high current values are to be used in the measurement, suitable pulse techniques may be used to provide pulses of short duty cycle to minimise the rise in junction temperature.

The following details are to be specified:

1 Test voltage or current.
2 Duty cycle and pulse width, if applicable.

Chapter 5 Measurements of circuit performance and thermal resistance

For thermal-resistance measurements, at least three temperature-sensitive parameters of the device can be used: the collector-to-base cut-off current (I_{CBO}), the forward voltage drop of the emitter-to-base diode (V_{EB}) and the forward voltage drop of the collector-to-base diode (V_{CB}). The methods described refer to the thermal resistance between specified reference points of the device. For this type of measurement, power is applied to the device at two values of case, ambient, or other reference point temperature, such that identical values of I_{CBO}, V_{EB} or V_{CB} are read during the cooling portion of the measurement.

Thermal resistance (collector cut-off current method)

This test measures the thermal resistance of the device under specified conditions and is particularly applicable to the measurement of devices having relatively large thermal response times.

Procedure

Switches S1 and S2 (Fig. 23) are ganged and operated such that the time they are closed (heat interval) is much larger than the time they are open (measurement interval). S1 is arranged to open slightly before S2 and the interval between the opening of S1 and S2 is adjusted to be short compared to the thermal time constant of the device being measured. When both switches are open, the value of I_{CBO} is read as the drop across R_B. If I_{CBO} varies during the measurement interval, the value immediately following the opening of S2 should be read. A calibrated oscilloscope makes a convenient detector. Care should be taken that the collector voltage stays constant.

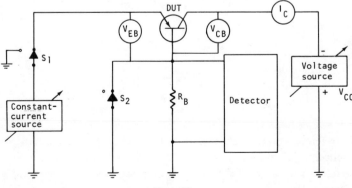

Figure 23

Measurement is made in the following manner. The case, ambient or other reference is elevated to a high temperature T_2, not exceeding the maximum junction temperature and the cut-off current, I_{CBO}, read with the constant-current source supplying no current. The reference temperature is then reduced to a lower temperature T_1, and power P_1 is applied to heat the device by increasing the current from the constant-current source, until the same value of I_{CBO} is read as was read above. Then

$$\theta = (T_2 - T_1)/P_1$$

where

$$P_1 = n(I_C V_{CC} + I_E V_{EB})$$

and n is the duty cycle (t_{on}/t_{total}).

Thermal resistance (forward voltage drop, emitter-to-base diode method)

This test measures the thermal resistance of the device under the specified conditions and is particularly applicable to the measurement of germanium and silicon devices having relatively short thermal response times.

Procedure

S1 (Fig. 24) is closed for a much larger interval (heat) than it is open (measurement). The measurement interval should be short compared with the thermal response time of the device being measured.

The constant measurement current is a small current (of the order of a few milliamperes) so selected that the magnitude of V_{BE} changes approximately 2 mV/degC of junction temperature change. The heating current source is adjustable.

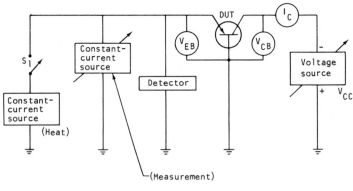

Figure 24

The measurement is made in the following manner. The case, ambient or other reference point is elevated to a high temperature T_2, not exceeding the maximum junction temperature and the emitter-to-base voltage V_{EB} is to be read at the start of the measurement interval. An oscilloscope makes a convenient detector.

At T_2 there will be a small power dissipated in the device due to the measurement current source. The reference temperature is then reduced to a lower temperature T_1 and power P_1 applied to heat the device by increasing the current from the constant current source, until the same value of V_{EB} is read as was read above. However, if P_1 is calculated as the heating power contributed by the heating current source only the equation

$$\theta = (T_2 - T_1)/P_1$$

gives accurately

$$P_1 = n(I_C V_{CC} + I_E V_{EB})$$

where n is the duty cycle (t_{on}/t_{total}).

Details

The following details are to be specified:

1 Test temperatures.
2 Test voltages or currents.

Thermal resistance (d.c. forward voltage drop, emitter-to-base, continuous method)

This test measures the thermal resistance of the device under specified conditions.

Procedure

The measurement technique assumes that the forward emitter voltage drop varies with temperature. It further assumes that during the course of measurement, the variation in forward emitter voltage drop varies monotonically due to temperature and is much greater than that due to the variation with collector voltage. The circuit used is shown in Fig. 25.

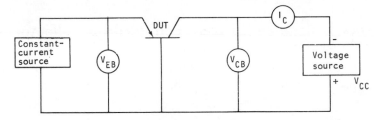

Figure 25

The measurement is made in the following manner. The case, ambient, or other reference point is elevated to a high temperature T_2, not exceeding the maximum junction temperature. Current I_C is set to a value and a voltage applied to the collector base diode, V_2. The value of V_2 applied should be low, yet high enough for the device to operate in a normal manner. V_{IEB} is read under these conditions. The reference temperature is reduced to a lower temperature T_1 and V_{cc} varied until the same value of V_{IEB} is read as was read above. The thermal resistance is then

$$\theta = (T_2 - T_1)/I_c (V_1 - V_2)$$

where V_1 is the collector voltage applied at temperature T_1.

Details

Details to be specified on the individual specification:
1. Test temperature.
2. I_c and V_2.

Thermal response time test

This test measures the time required for the junction to reach 90% of the final value of junction temperature change following the application of a step function of power dissipation. The apparatus used to determine the thermal response time should be capable of demonstrating device conformance to the minimum requirements of the individual specification.

Procedure

The thermal response time is to be determined by measuring the time required for the junction (as indicated by a precalibrated temperature-sensitive electrical parameter) to reach 90% of the final value of the junction temperature change caused by a step function in power dissipation when the case or ambient temperature, as specified, is held constant.

Details

The following detail should be specified:
1. Device case or ambient temperature.

Thermal resistance (forward voltage drop, collector-to-base diode method)

This test method is particularly applicable to the measurement of germanium and silicon devices having relatively long thermal response times.

Procedure

Switches S1 and S2 (Fig. 26) are ganged switches and are so arranged that S2 very shortly after S1 opens and such that the delay between the openings is much shorter than the thermal response time of the device being measured. S1 and S2 should be closed (heat interval) for a much larger time than they are open (measurement interval) and the measurement interval should be short compared with the thermal response time of the device being measured.

The measurement is made in the following manner. The case, ambient,

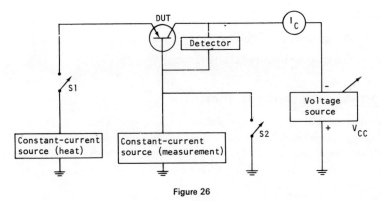

Figure 26

or other reference point is elevated to a high temperature T_2 not exceeding the maximum junction temperature and the collector-base voltage, V_{CB}, is read. This reading is made at the beginning of the measurement interval. An oscilloscope makes a convenient detector.

The reference temperature is then reduced to a lower temperature, T_1. The heating power, P_1, is adjusted by adjusting the heating current source in the emitter circuit until the same value of V_{CB} is read as was read above. The value of θ is then given by

$$\theta = (T_2 - T_1)/P_1$$

where

$$P_1 = n(I_C V_{CC} + I_E V_{EB})$$

and n the duty cycle (t_{on}/t_{total}).

Details

The following details are to be specified:

1 Test temperatures.
2 Test voltages and currents.

Thermal time constant test

This term measures the time required for the junction to reach 63·2% of the final value of junction temperature change following application of a step function of power dissipation. The apparatus used to determine the conformance to the minimum requirements.

Procedure

The thermal time constant should be determined by measuring the time required for the junction temperature (as indicated by a precalibrated temperature sensitive electrical parameter) to reach 63·2% of the final value of junction temperature change caused by a step function in power dissipation, when the device case or ambient temperature is held constant.

Details

The following detail is to be specified on the individual specifications:

1 Device case or ambient temperature.

Thermal resistance (general)

This test measures the temperature rise per unit power dissipation of a designated junction above the case or ambient temperature, under conditions of steady-state operation. The apparatus used to determine the thermal resistance should be capable of demonstrating device conformance to a minimum requirement.

Procedure

The thermal resistance may be determined by:

1 Measuring the junction power required to maintain the junction temperature constant (as indicated by a precalibrated temperature-sensitive electrical parameter) when the case or ambient temperature is changed by a known amount.
2 Measuring the junction temperature (as indicated by a precalibrated temperature-sensitive electrical parameter) when the junction power is changed by a known amount while the case or ambient temperature is held constant.

Details

The following detail is to be specified:

1 Characteristic being measured θ_{J-C} or θ_{J-A}.

Thermal resistance (d.c. current gain, continuous method)

This test measures the thermal resistance of the device.

Procedure

The measurement technique assumes that the current gain varies with temperature. The rate of change is unimportant. It further assumes that during the course of measurement the variation in current gain varies monotonically due to temperature variation and is much greater than that due to the variation with collector voltage. The circuit is shown in Fig. 27.

The measurement is made in the following manner. The case, ambient, or other reference point is elevated to a high temperature T_2 not exceeding the maximum junction temperature. Current I_C is set to a value and a voltage applied to the collector base diode. V_2. The value of V_2 applied should be low, yet high enough for the device to operate in a normal manner.

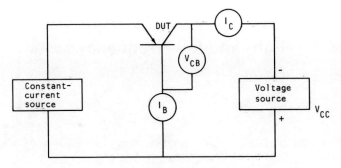

Figure 27

I_B is read under these conditions. The reference temperature is reduced to a lower temperature T_1 and V_{CC} varied until the same value I_B is read as was read above. The thermal resistance is then

$$\theta = (T_2 - T_1)/I_C(V_1 - V_2)$$

where V_1 is the collector voltage applied at temperature T_1.

Details

The following details are to be specified:

1. Test temperature.
2. I_C and V_2.

Chapter 6 High- and low-frequency tests

Unless otherwise specified, the measurement should be made at the electrical test frequency, 1 kHz ± 25%. At this frequency the reactive components may not be negligible.

Small-signal short-circuit input impedance test

The purpose of this test is to measure the input impedance of the device. The circuit and procedure used are for the common emitter configuration. For other parameters the circuit should be changed accordingly.

The biasing circuit shown in Fig. 28 is solely for the purpose of illustration—other stable biasing circuits may be used.

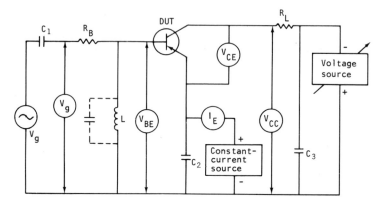

Figure 28

Procedure

The capacitors C_1, C_2 and C_3 should present short-circuits at the test frequency in order to effectively couple and by-pass the test signal. The inductance L should be resonated with a capacitor and the combination should have a large impedance compared with h_{ie} at the test frequency. R_L should be a short-circuit compared with the output impedance of the device. V_g and V_{be} are measured on high-impedance a.c. voltmeters after setting the specified values I_E and V_{CE}. Then

$$h_{ie} = V_{be}/I_b$$

Details

Details to be specified:
1. Test frequency.
2. Test voltages and currents.
3. Parameter to be measured.

Small-signal open-circuit admittance

This test measures the output admittance of a device. The circuit and test procedure are for the common emitter configuration. For other parameters the circuit and procedure should be changed accordingly. The biasing circuit shown (Fig. 29) is solely for the purpose of illustration—other stable biasing circuits may be used.

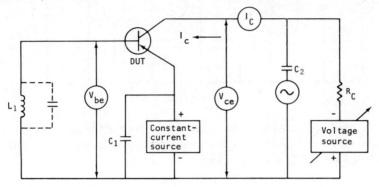

Figure 29

Procedure

Inductance L_1 should be resonated with a capacitor and the combination should have a large impedance compared with h_{ie} at the test frequency. C_1 and C_2 should present short-circuits at the test frequency in order to effectively couple and by-pass the test signal. Voltmeters V_{BC} and V_{CE} should be high-impedance voltmeters. Then

$$h_{oe} = I_c/V_{ce}$$

Details

Details to be specified:
1. Test voltages and currents.
2. Test frequency.
3. Parameter to be measured.

Small-signal short-circuit forward-current transfer ratio test

This test measures the forward-current transfer ratio of a device. The circuit and procedure shown are for the common emitter configuration. For other parameters the circuit and the procedure should be changed accordingly. The biasing circuit shown (Fig. 30) is solely for the purpose of illustration—other stable biasing circuits may be used.

Procedure

The capacitors C_1, C_2 and C_3 should present short-circuits at the test frequency in order to effectively couple and by-pass the test signal. The inductance L should be resonated with a capacitor and the combination should have a large impedance compared with h_{ie} at the test frequency. R_L should be a

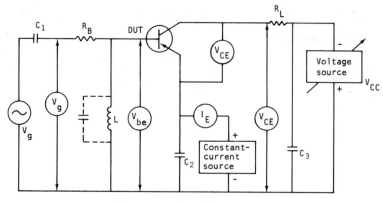

Figure 30

short-circuit compared with the output impedance of the device. V_g, V_{be} and V_{ce} should be measured on high-impedance a.c. voltmeters after setting the specified values of I_E and V_{CE}. Then

$$h_{fe} = I_c/I_b$$

where

$$I_c = V_{ce}/R_L$$

Details

The following details are to be specified:
1. Test frequency.
2. Test voltage and currents.
3. Parameter to be measured.

Small-signal open-circuit reverse-voltage transfer ratio test

This test measures the reverse-voltage transfer ratio of the device. The circuit and procedures shown are for the common emitter configuration. For other parameters the circuit and procedures should be changed accordingly. The biasing circuit shown (Fig. 31) is solely for the purpose of illustration—other stable biasing circuits may be used.

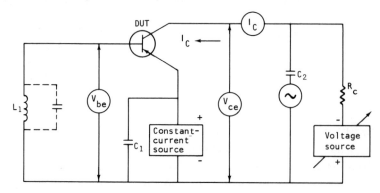

Figure 31

Procedure

Inductance L_1 should be resonated with a capacitor and the combination should have a large impedance compared with h_{ie} at the test frequency. Capacitors C_1 and C_2 should present short-circuits at the test frequency in order to effectively couple and by-pass the test signal. Voltmeters V_{be} and V_{ce} should be high-impedance voltmeters. Thus, after applying the specified test voltages and currents,

$$h_{re} = V_{be}/V_{cr}$$

Details

Details to be specified:

1. Test frequency.
2. Test voltages and currents.
3. Parameter to be measured.

Small-signal short-circuit input admittance test

This test is to measure the input admittance of the device. The circuit and procedure shown are for the common emitter. For other parameters both the circuit and the procedure should be changed accordingly. The biasing circuit shown (Fig. 32) is solely for the purpose of illustration—other stable biasing circuits may be used.

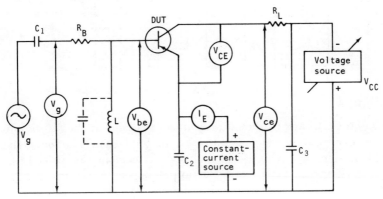

Figure 32

Procedure

Capacitors C_1, C_2 and C_3 should present short-circuits at the test frequency in order to effectively couple and by-pass the test signal. The inductance L should be resonated with a capacitor and the combination should have a large impedance compared with h_{ie} at the test frequency. R_L is optional and should be a short-circuit compared with the output impedance of the device. V_2 and V_{be} are measured on high-impedance a.c. voltmeters. Then

$$h_{ie} = V_{be}/I_b$$

Thus

$$Y_{ie} = 1/h_{ie}$$

Details

The following details should be specified:

1. Test frequency.
2. Test voltages and currents.
3. Parameter to be measured.

Small-signal short-circuit forward-transfer admittance test

The purpose of this test is to measure the forward transfer admittance of the device. The circuit and procedure are for the common emitter. For other parameters both circuit and procedure should be changed accordingly. The biasing circuit shown (Fig. 33) is solely for the purpose of illustration—other stable biasing circuits may be used.

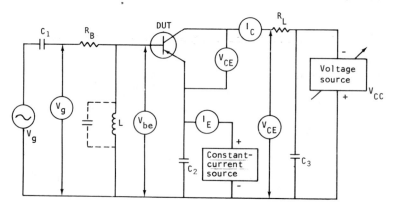

Figure 33

Procedure

Capacitors C_1, C_2 and C_3 should present short-circuits at the test frequency in order to effectively couple and by-pass the test signal. The inductance L should be resonated with a capacitor and the combination should have a large impedance compared with the output impedance of the device. V_g, V_{be} and V_{ce} should be measured using high-impedance a.c. voltmeters. Then

where
$$Y_{fe} = I_c/V_{be}$$

Thus
$$I_c = V_{ce}/R_L$$

$$Y_{fe} = (V_{ce}/V_{be})(1/R_L)$$

Details

The following details are to be specified:

1. Test frequency.
2. Test voltages and currents.
3. Parameter to be measured.

Small-signal short-circuit reverse-transfer admittance test

This test measures the reverse transfer admittance of a device. The test circuit is shown in Fig. 34, but for other configurations the circuit should be modified in such a manner that the circuit is capable of demonstrating the device conformance to the specification.

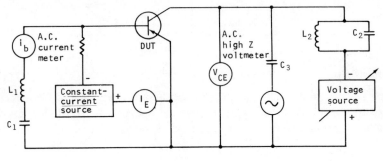

Figure 34

Procedure

The capacitor C_3 should effectively present a short-circuit to the test signals, L and C should be made to resonate at the test frequency to effectively present an open-circuit to the test signal.

The bias voltages and current specified should be applied to the terminals; the input terminals to be a.c. short-circuit. The magnitude, without regard to phase angle, of the ratio of input current and output voltage should then be measured. Then, for the common emitter

$$Y_{re} = i_b/V_{ce}$$

for the common collector

$$Y_{rc} = i_b/V_{ec}$$

and for the common base

$$Y_{rb} = i_e/V_{ab}$$

Details

Details to be specified:

1. Test bias voltage and current.
2. Test frequency, if different from 1 KHz.

Small-signal short-circuit output admittance test

This test measures the output admittance of the device under certain conditions. The circuit (Fig. 35) and procedure shown are for the common emitter configuration. For other parameters the circuit and procedure should be amended accordingly.

70 Semiconductor devices

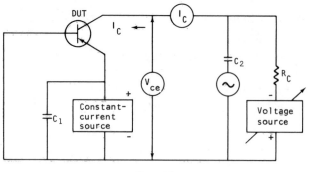

Figure 35

Procedure

Capacitors C_1 and C_2 should present short-circuits at the test frequency in order to effectively couple and by-pass the test signal. Resistance R_c is not zero but chosen for any convenient value. Then

$$Y_{oe} = I_c/V_{ce}$$

Details

The following details should be specified:

1 Test frequency.
2 Test voltages and currents.
3 Parameter to be measured.

Open-circuit output capacitance test

This test is designed to measure the open circuit capacitance of the device. The circuit (Fig. 36) and procedure shown are for the common base configuration. Other parameters can be measured if the circuit and procedure are modified.

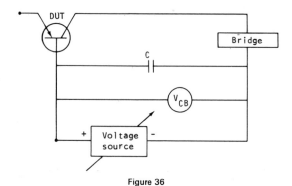

Figure 36

Procedure

The bridge should have a low d.c. resistance between its output terminals and be capable of carrying the specified collector current without affecting the desired accuracy of measurement. The emitter should be open-circuited to a.c. and the frequency of measurement as specified. Capacitor C should be sufficiently large to provide a short-circuit at the test frequency.

The capacitance-reading instrument is nulled with the circuitry connected, thereby eliminating errors due to stray capacitance of the circuit wiring.

The device to be measured is inserted into the test socket and properly biased; the output capacitance is then measured.

Details

The following details should be stated:

1. Test voltages or currents.
2. Measurement of frequency.
3. Parameter to be measured.

Input capacitance test

The purpose of this test is to measure the shunt capacitance of the input terminals of the device. The circuit is shown in Fig. 37, but for other configurations the circuit may be modified in such a manner that is capable of demonstrating device conformance to the minimum requirements.

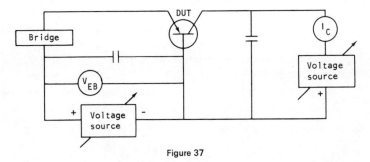

Figure 37

Procedure

The bridge should have a low d.c. resistance between the input terminals and be capable of carrying the required emitter current without affecting the desired accuracy of measurement. The specified voltages and current should be applied to the terminals; an a.c. small signal should be applied to the input terminals and the output terminals to be a.c. short circuited. The input capacitance is then measured. The capacitance reading instrument is nulled with the circuitry connected, thereby eliminating errors due to stray capacitance and circuit wiring.

Details

Details to be specified are:

1. Test voltages or currents.
2. Test frequency.

Direct inter-terminal capacitance test

This test measures the direct terminal capacitance between terminals using electrical biases.

Procedure

A direct capacitance or resonance method can be used to determine the value of the direct inter-terminal capacitance.

Method 1

The specified voltage should be applied between specified terminals; an a.c. small signal is applied to the terminals and the direct inter-terminal capacitance measured. The lead capacitance beyond 12·5 mm from the body seat is to be effectively eliminated by suitable means such as a test socket shielding. The abbreviations and symbols used are defined as follows:

$C_{cb(dir)}$—collector-to-base inter-terminal direct capacitance.
$C_{eb(dir)}$—emittar-to-base inter-terminal direct capacitance.
$C_{ce(dir)}$—collector-to-emitter inter-terminal direct capacitance.

Method 2

A suitable resonance method can be used to measure the following two terminal capacitances:

C_1—capacitance between collector terminal and ground, with base and emitter terminals grounded.
C_2—capacitance between base terminal and ground, with collector and emitter terminals grounded.
C_3—capacitance between collector and base terminals strapped together and ground, with the emitter terminal grounded.

The direct inter-terminal capacitance can then be calculated from

$$C_{cb(dir)} = \tfrac{1}{2}(C_1 + C_2 - C_3)$$

The direct inter-terminal capacitance for other configurations can be determined by suitable modifications of the above procedure. Such modifications should be capable of demonstrating device conformance to the minimum requirements.

Details

The following details are to be specified:
1. Terminal arrangement.
2. D.c. biasing conditions.
3. Test voltage or current.
4. Measurement frequency.

Diffusion capacitance test

This test measures the diffusion capacitance of the device.

Procedure

The bridge used in the circuit (Fig. 38) is to be capable of carrying the required emitter current low (d.c. resistance between its output terminals) without affecting the desired accuracy of measurement. In addition, the bridge should be capable of demonstrating conformance of the specification. The capacitance reading instrument is nulled with the circuit connected, thereby eliminating errors due to stray capacitance of the circuit wiring.

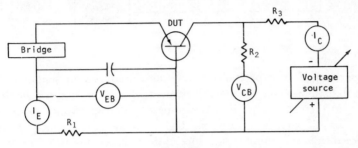

Figure 38

The forward current or forward voltage should be applied to the terminals. An a.c. small signal is applied to the emitter-base terminals and the collector-base terminals open-circuited to a.c. The specified shunt capacitance is then measured. A representative device encapsulation, without device element, should be inserted in the test socket and the shunt capacitance measured. The capacitance so measured is the parasitic capacitance of the device encapsulation which should be subtracted from the shunt capacitance of the device tested in order to determine the diffusion capacitance.

Details

The following should be specified:
1. Test voltage and current.
2. Test frequency.

Depletion-layer capacitance test

This test measures the depletion-layer capacitance of the device.

Procedure

The bridge used in the circuit shown in Fig. 39 should be capable of carrying the required collector current without affecting the desired accuracy of measurement. In addition, the bridge should be capable of demonstrating conformance to the requirements of the individual specification. The capacitance reading instrument is nulled with the circuitry connected, to eliminate errors due to the stray capacitance of the circuit wiring.

The specified reverse voltage or reverse voltage and current should be applied to the terminals. An a.c. small signal is to be applied to the collector-base terminals and the emitter-base terminals to be open-circuited to a.c.

74 Semiconductor devices

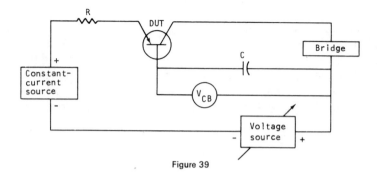

Figure 39

The specified shunt capacitance is now measured. A representative device encapsulation, without the device element, should be inserted in the test socket and the shunt capacitance measured. The capacitance so measured is the parasitic capacitance of the device encapsulation which should be subtracted from the shunt capacitance of the device being tested in order to determine the depletion-layer capacitance.

Details

Details to be specified:

1. Test current and voltage.
2. Test frequency.

Noise figure test

This test is used to measure the noise figure of the device. An average-responding r.m.s. calibrated indicator should be used in addition to other suitable apparatus to measure the noise factor of the device.

Procedure

The voltage and current specified should be applied to the terminals and the noise figure measured at the frequency specified (normally 1000 Hz) with an input resistance of 1,000 Ω and as referred to a one cycle bandwidth.

Details

The following details are to be specified:

1. Test voltage and current.
2. Test frequency.
3. Load resistance.

Pulse response test

This test is used to measure the pulse response of the device.

Procedure

The pulse response of the device should be measured using test condition 1 or 2.

Test condition 1
The device should be operated in the common emitter configuration as shown in Fig. 40a with the collector load resistance R_c and the collector supply voltage specified. When measuring the delay or rise time, $I_{B(0)}$ and $I_{B(1)}$ or $V_{BE(1)}$ should be specified. The input transition and the collector

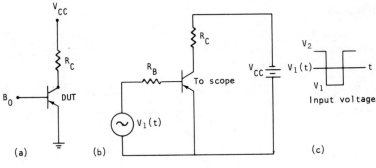

Figure 40

voltage response detector should have rise and response fall times such that doubling these responses will not affect the results greater than the precision of measurement. The current and voltages specified should be constant. Stray capacitance of the circuit should be sufficiently small so that doubling it does not affect the test results greater than the precision of measurement.

$I_{B(0)}$ —Prior off state base current
$V_{BE(0)}$ —Prior off state base to emitter voltage
$I_{B(1)}$ —On state base current
$V_{BE(1)}$ —On state base to emitter voltage
$I_{B(2)}$ —Post off state base current
$V_{BE(2)}$ —Post off state base to emitter voltage.

Test condition 2
The device should be operated in the test circuit shown in Fig. 40b (constant-current drive) with the voltages and component values as specified. The pulse or square wave generator and scope should have rise and fall response times such that doubling these responses will not affect the results greater than the precision of measurement.

Details

The following details are to be specified:

1. Test condition (1 or 2).
2. Collector load resistance (R_c) and the collector supply voltage (V_{cc}) for method 1.

76 Semiconductor devices

3 Base resistance (R_B), collector load resistor (R_C), and collector supply voltage (V_{cc}) for method 2.
4 Test voltages or current.

Small-signal power gain test

The purpose of this test is to measure the ratio of the a.c. output power to the a.c. input power (usually specified in decibels), for small-signal gain.

Procedure

For other configurations the circuit shown in Fig. 41 should be modified in such a manner that the circuit is capable of demonstrating device performance to the specification.

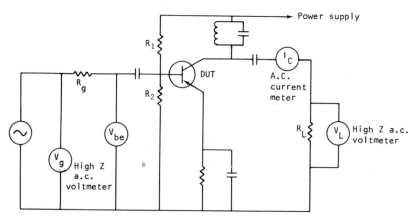

Figure 41

The specified voltage(s) and or current(s) should be applied to the input terminals; an a.c. small signal should be applied to the input terminals of the circuit. Resistances R_1 and R_2 should have values larger than h_{ie} of the device. The phase angle θ between the input current and V_{be} should be considered to be 0 if the specified test frequency is less than the extrapolated unity frequency (f_t) of the device. Then for common emitter:

$$P_{ge} = 10 \log (P_{out}/P_{in})$$

where

$$P_{in} = V_{be} i_b \cos \theta$$
$$i_b = (V_g - V_{be})/R_g$$
$$P_{out} = i_c^2 R_L \text{ or } V_L^2/R_L$$

Thus

$$P_{ge} = 10 \log \left[\frac{i_c^2 R_L}{V_{be}(V_g - V_{be})/R_g} \right]$$

or

$$P_{ge} = 10 \log \left[\frac{V_L^2/R_L}{V_{be}(V_g - V_{be})/R_g} \right]$$

High- and low-frequency tests

For other configurations, only modifications capable of demonstrating device conformance to the specification should be made.

Details

The following details should be specified:

1. Test circuit.
2. Test voltage (2) and/or current(s).
3. Test frequency.

Extrapolated unity gain frequency test

This test determines the extrapolated unity gain frequency of the device.

Test circuit

The test circuit employed in determining the extrapolated unity gain frequency is that which is used for measuring the magnitude of the common emitter small-signal short-circuit current transfer ratio (see Fig. 43).

Procedure

The magnitude of the common emitter short-circuit current transfer ratio should be determined at the specified signal frequency with specified bias voltages and/or currents applied. The product of the specified signal frequency (f) and the measured common emitter small-signal short-circuit transfer ratio (h_{fe}) is the extrapolated unity gain frequency (f_t).

Details

Details to be specified:

1. Test current and voltage.
2. Test frequency.

Real part of small-signal short-circuit input impedance test

This test measures the resistive component of the small-signal short-circuit impedance of the device.

Procedure

The circuit shown in Fig. 42 is used for measuring the common emitter real part of the small-signal short-circuit input impedance. For other device configurations, the above circuitry should be modified so that it is capable of demonstrating conformance to minimum requirements.

78 Semiconductor devices

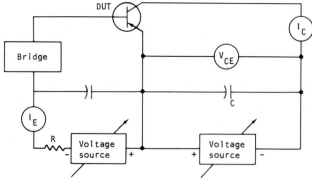

Figure 42

Details

Details to be specified:

1. Test voltage and current.
2. Test frequency.

Small-signal short-circuit forward-current transfer ratio

This test measures the forward-current transfer ratio cut-off frequency.

Test circuit

The circuit and procedure shown in Fig. 43 are for the common base configuration. For other parameters the circuits and procedure should be changed accordingly. Normal VHF circuit precautions should be taken. At frequencies higher than 10 mHz the use of this circuit may lead to excessive errors. The biasing circuit shown is for the purpose of illustration only—any biasing circuit may be used.

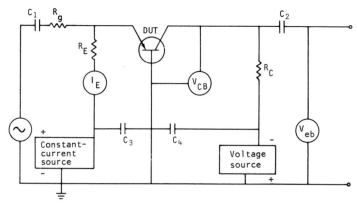

Figure 43

Procedure

The voltages and currents should be specified. Resistances R_G and R_E are to be large so as to present open circuits to h_{ib}. Resistance R_c should be small to present a short circuit to h_{ob}. Capacitors C_1, C_2 and C_3 should present short-circuits at the test frequency to effectively couple and by-pass the test signal.

The circuitry is to be frequency-independent. This can be checked by removing the device from the circuit and shorting between emitter and collector with no bias voltages applied. Care should be taken to ensure that the generator has a sufficiently pure waveform and that the high-impedance voltmeter is adequately sensitive to enable the measurements to be made at a low enough signal level to avoid the introduction of harmonics by the device.

The generator is set to a frequency at least 30 times lower than the lowest cut-off frequency limit and the low frequency h_{fb} is measured. The frequency is then increased until the magnitude of h_{fb} has fallen to $1/\sqrt{2}$ of its low frequency value. This is the cut-off frequency.

Details

Details to be specified:

1. Test voltages and currents.
2. Parameter to be measured.

Small-signal short-circuit forward-current transfer ratio test

This test is to measure the forward-current transfer ratio.

Test circuit

The circuit (Fig. 44) and procedure shown are for the common emitter configuration. For other parameters the circuit and procedure should be changed accordingly. The biasing circuit shown is solely for the purpose of illustration—other stable biasing circuits may be used.

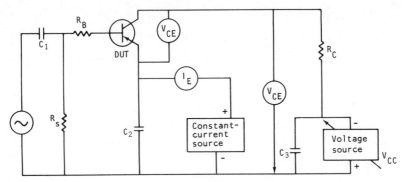

Figure 44

Procedure

Capacitors C_1, C_2 and C_3 should present short-circuits in order to effectively couple and by-pass the test signal at the frequency of measurement. The value R_B should be sufficiently large to provide a constant current source.

Resistor R_C should be a short-circuit compared to the output impedance of the device. With the device removed from the circuit, a shorting link is placed between the base and collector and the output voltage of the signal generator is adjusted until a reading of one (in arbitrary units) is obtained on the high-impedance a.c. voltmeter V_{cc}. With the device in the circuit and biased as specified, the reading on the voltmeter V_{cc} is now equal to the magnitude of (h_{fe}).

Details

The following details are to be specified:
1. Measurement frequency.
2. Test voltages and currents.
3. Parameter to be measured.

Chapter 7 General diode tests

Capacitance test

This test is designed to measure the capacitance of the device. The circuit used is shown in Fig. 45.

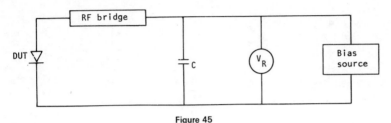

Figure 45

Procedure

The RF bridge should be capable of carrying the current without affecting the accuracy of the measurement. Small-signal conditions should prevail during the measurement. The specified voltage is to be applied and the capacitance measured at the specified frequency.

Details

The following details should be stated:

1. Test voltage.
2. Test frequency.

D.C. output current test

This test measures the average output, rectified current of the device. The circuit used is shown in Fig. 46.

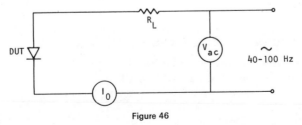

Figure 46

Procedure

The specified a.c. input voltage is applied at any convenient frequency within the range of 40 to 100 Hz. The load resistance R_L should be specified.

Details

The following are to be specified:

1. Test voltage.
2. Load resistor.

Forward current and forward voltage test

This test is designed to measure the voltage and current, in the forward direction, through the device. The circuit is shown in Fig. 47

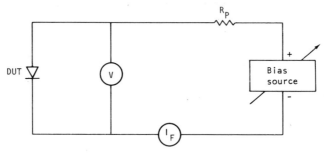

Figure 47

Procedure

Forward current

The specified forward voltage is to be applied to the terminal of the device and the forward current measured. The voltmeter used to measure V_F should present an open-circuit to the device.

Forward voltage

The supply voltage is adjusted to obtain the specified value of the forward current through the device. The forward voltage is then read from the voltmeter.

Details

The following details are required:

1. Test voltage.
2. Test current.

Reverse current and reverse voltage test

This test measures the voltage and current, in the reverse direction, through the device. The current is shown in Fig. 48.

Procedure

Reverse current

The supply voltage is adjusted to obtain the specified value of reverse voltage across the device. The reverse current is then read from the current meter.

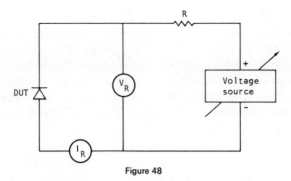

Figure 48

Reverse voltage

The reverse current should be adjusted from zero to the specified value and the reverse voltage measured. The maximum voltage rating of the device should not be exceeded. The test voltage and current should be specified.

Breakdown voltage test

This test measures the voltage across the device in the breakdown region.

Test circuit

The voltmeter being used to measure the terminal voltage should present an open-circuit to the terminals across which the voltage is being measured (see Fig. 49). The ammeter should present essentially a short-circuit to the terminals between which the current is being measured.

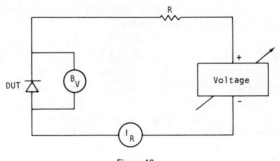

Figure 49

Procedure

The reverse current is to be adjusted from zero until either the maximum voltage or the specified maximum current is reached. The device is acceptable if the specified maximum limit for B_V is reached before the current reaches the specified value. If the specified current is reached first, the device is rejected. The test current should be specified.

Forward recovery time test

This test is intended to measure the forward recovery time of the device. A device reveals an excessive transient voltage drop when it is switched rapidly into the forward conductance region. The amplitude or time duraction of this voltage peak is of importance in many applications of devices. The amplitude or time duration of this voltage peak can be measured by observing the voltage waveform across the device when a square-wave signal of the specified amplitude, rise time and pulse width is applied to the device. The circuit and method of measurement described here have been selected to yield the most repeatable results with a wide variety of input and output equipments. The test circuit is shown in Fig. 50.

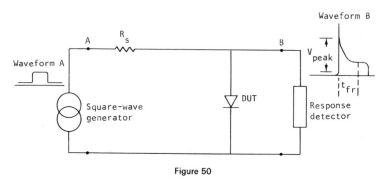

Figure 50

Procedure

The following conditions and limits for square waves should prevail to achieve the utmost standardisation of results:

1. Amplitude = 1, 2, 5 and 10^n (where n is an integer).
2. Rise time = $0 \cdot 1x$ or less specified response time or as stated (10 to 90%).
3. Pulse width ($50 = 10x$ or more specified response time or as stated percentage of maximum duty cycle).
4. Source resistance = $20R_F$ or more ($R_F = V_F/I_F$ at specified I_F), e.g. doubling the source resistance should not change the output reading by more than the required precision of measurement.
5. Load impedance z (halving = $100R_F$ or more, the load impedance, or changing the load resistor by a factor of 2, should not affect the output reading by more than the required precision of measurement).
6. Limits: forward voltage (instantaneous) = V_{fr} and the forward recovery time is t_{fr} measured from the instant of pulse ascension.

Details

The test conditions and limits are to be specified.

Reverse recovery time test

The purpose of this test is to measure reverse recovery time by observing the reverse transient current through a specified load resistance on switching from a specified forward bias to a specified reverse bias. For diodes exhibiting reverse recovery times greater than or equal to 300 nsec, the following procedure should be used.

General diode tests 85

Test circuits

The test circuit is shown in Fig. 51 (and in simplified form in Fig. 52).

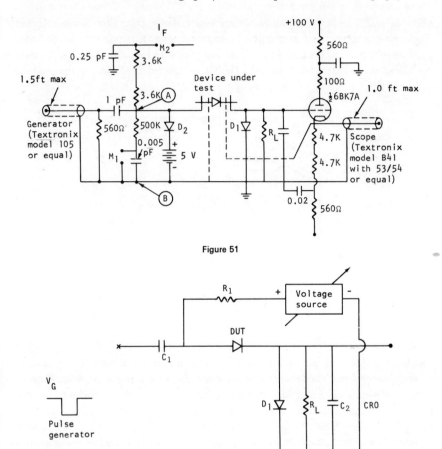

Figure 51

Figure 52

Procedure

Resistor R_1 should be of a high resistance. The forward current is set to the specified value. Resistor R_L is the load resistor and is shunted by a diode D_1 which should be faster than the device under test, in order to prevent over-driving the CRO amplifiers. When it is not feasible to find a suitably fast diode by direct measurement of its recovery characteristics the following method should be used.

Pick at random a minimum of 100 diodes of the fastest type available which will meet the d.c. requirements. Then using any reasonably fast diode as a reference (it should meet the d.c. requirements, and the specified test conditions), select from a group of 100 the diode that appears to be the fastest. In doing this certain diodes will probably cause the trace to swing above the zero reference. This is normal and desirable, since it indicates that such diodes are faster than the reference diode being used (see Fig. 53).

The fastest diode is the one that has the biggest net area above the reference line. Replace the reference diode with the fastest one selected from the lot of 100 and again read the remaining 99. Of these pick the fastest 20. If none of these 20 is faster than the reference diode (the first diode selected from the group of 100) then go on to the next step; however, if one or more of the 20 appear to be faster it will be necessary to use that diode as a reference and reread the other 20 until the single fastest diode in the original lot of 100 is found.

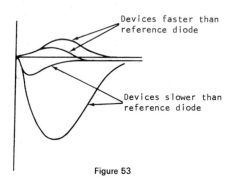

Figure 53

Now test the remaining 20 diodes and record their readings using the fastest diode as reference. Also select from the group to be tested a diode as close to the limit as possible. Then read the limit diode using as reference a diode from the group of 20 which has a recorded reading of at least 5% of the limit value and no more than 10% of the limit value. If the second reading of the limit diode varies by no more than 10% from the first reading, the fastest diode of the group of 100 can be used as a reference diode for the test in question. Otherwise a faster diode should be found by sorting a larger group of diodes. The 20 diodes with recorded readings should be kept as a control lot to check the selected reference. Then, at any time these diodes can be reread if their readings vary appreciably from the recorded value (and the rest of the equipment is in proper calibration) the reference diode has changed and should be replaced.

In considering the role of the reference diode, it is well to realise that diodes being tested should be 10 to 20 times slower than the reference for the equipment to give absolute readings (correlation can be obtained with less differences).

The procedure described for selecting a reference ensures that this is the case for limit or slower diodes, the critical ones in a production test—and makes it possible to select such a reference out of production.

If absolute readings are desired at a lower current than the test limits, that lower current should be considered the limit when selecting a reference.

Details

The following details should be specified:
    1 Forward voltage.

2 Reverse current.
3 Load resistance.
4 Load capacitance.

For diodes exhibiting reverse recovery time less than 300 nsec, the following procedure should be used:

Test circuit

The circuits used are shown in Figs 54 and 55.

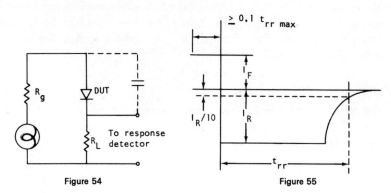

Figure 54   Figure 55

Procedure

The diode should be switched in a circuit having an RC time constant including the diode capacitance which is less than (normally $0 \cdot 1 t_{rr(max)}$) or equal to that specified. The specified forward current should be applied for a minimum duration of ten times the maximum reverse recovery time specified prior to switching to the specified reverse current. The diode should be considered to recover when the reversed current is reduced to one-tenth of the specified initial value. The current transition and the response time should be such that doubling these responses will not affect the required precision of measurement. The specified current and voltages should be constant.

Details

The following details should be specified:
1 Forward current.
2 Reverse current.
3 Load resistance.
4 Load capacitance.
5 RC time constant.

'Q' for variable capacitance diodes test

This test measures the quality factor (or Q-factor) of the device.

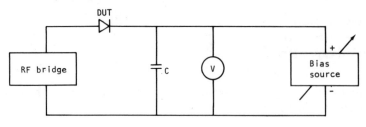

Figure 56

Procedure

The parallel resistance, R_p, and capacitance C_p, of the device (see Fig. 56) should be measured by RF bridge methods under specified conditions of frequency and bias. The capacitor C should be a short-circuit at the measurement frequency in order to effectively by-pass the RF signal. Small-signal conditions should be used. Then

$$Q = 2\pi f R_p C_p$$

Details

The test frequency and test voltages and currents should be specified.

Rectification efficiency test

This test measures the rectification efficiency of the device.

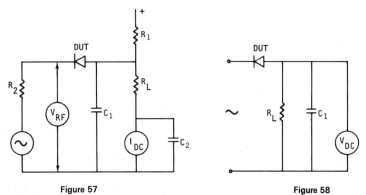

Figure 57   Figure 58

Procedure

Test condition 1

Resistor R_L and capacitor C_1 comprise the load circuit and should be specified. Resistor R_1 in conjunction with R_L provides the specified bias current for the output current meter I_{DC}. The change in output current Δ d.c. is measured when the a.c. signal is applied (Fig. 57). The frequency and amplitude impedance of the generator should be specified.

Test condition 2

An alternative method uses a voltmeter across the load circuit. The voltmeter should preferably be a high impedance voltmeter (Fig. 58). Then

$$\text{Rectification efficiency} = (V_{DC}/V_{r.m.s.})\ 100\%$$

where $V_{r.m.s.}$ is the r.m.s. input.

Details

The following details should be specified:

1. Test condition letter.
2. Load circuit details.
3. Bias current.
4. Frequency and amplitude of signal.
5. Generator impedance.

Average reverse current test

This test is designed to measure the average reverse current through the device.

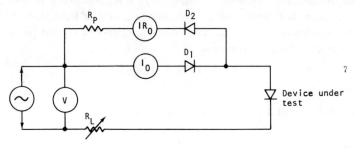

Figure 59

Test circuit

The reverse leakage current at each device, D1 and D2 must be less than 1/20th of the maximum allowable specified leakage current of the device under test. In other respects, the D1 and D2 should be the same type as the device being tested.

Procedure

After thermal equilibrium, at the temperature specified, the specified voltage should be applied.

Details

The test temperature as well as the test voltage is to be specified.

Small-signal breakdown impedance test

This test measures the breakdown impedance of the device.

Procedure

The bias supply to the device must present an a.c. open circuit across the device (Fig. 60), i.e. resistance R_1 should be large compared with the breakdown impedance being measured. The specified breakdown current or voltage should be applied using current meter I_R or voltmeter V_R.

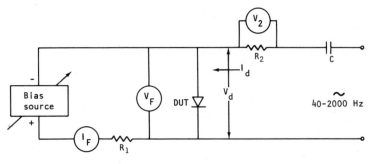

Figure 60

An a.c. signal in the frequency range 40-2,000 Hz should be applied through an isolating capacitor C and a current measurement resistor R_2.

The r.m.s. signal should not be greater than 10% of the value of the d.c. reverse breakdown current, I_R. The voltages V_d and V_2 should be measured using high-impedance voltmeters. Then

$$Z_{br} = V_d/I_d = V_d R_2/V_2$$

Details

The test voltages and currents are to be specified.

Surge current test

The purpose of this test is to determine the ability of the device to withstand current or voltage surges.

Procedure

The d.c. current or voltage should be specified and applied to the device. A specified number of current pulses should be superimposed on the average device current at a low duty cycle. The pulses should be half-sine waveform for a given duration. The duty cycle to be chosen so that the junction temperature is not changed significantly.

Details

The following details are to be specified:
1 Test voltages or currents.
2 Number of current pulses.
3 Duration of pulses.
4 Duty cycle of pulses.

Temperature coefficient of breakdown voltage test

This test measures the temperature coefficient of breakdown voltage. The apparatus used to measure the temperature coefficient should be capable of demonstrating device conformance to the minimum specified requirement of the device.

Procedure

The temperature coefficient of breakdown voltage is the percentage deviation of the voltage from the breakdown voltage obtained at the specified test temperature.

Details

Details required are:
1. Temperature.
2. Test current.

Small-signal forward impedance test

This test measures the forward impedance of the device under small-signal conditions.

Procedure

The bias supply to the device should present an a.c. open-circuit across the device, i.e. R_1 should be large compared with the forward impedance being measured (see Fig. 61).

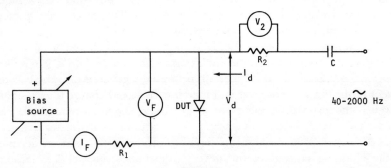

Figure 61

The specified voltage or current should be applied and monitored by the current meter I_F and voltmeter V_F. An a.c. signal in the frequency range 40–2,000 Hz should be applied through an isolating capacitor C and the current measurement resistor R_2.

The r.m.s. signal should not be greater than 10% of the value of the d.c. forward current I_F. The voltage V_d and V_2 are to be measured using high impedance voltmeters. Then:

$$Z_f = V_d/I_d = V_d R_2/V_2$$

Details

The test voltages or current are to be specified.

Stored charge test

The purpose of this test is to measure the charge-storage properties of the device.

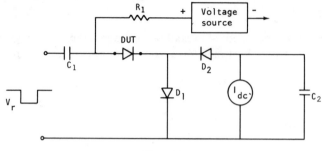

Figure 62

Procedure

Resistor R_1, which supplies the specified forward current to the device under test, should be of high resistance.

The time constant $R_1 C_1$, should be long compared with the pulse width, but C_1 should be small enough to be fully discharged during the interval between pulses. The pulse generator should have an output impedance of less than 100 Ω; the rise time of the pulse should be short enough and the pulse length large enough, that further change will not alter the accuracy of measurement. Device D1 passes the forward current of the device under test. It should have a high resistance in the reverse direction and a much smaller stored charge than the device under test. Device D2, has a small reverse bias during the interval between pulses and its leakage current should be small. During the pulse, D2 passes the stored charge of the device under test to capacitor C_2 and to the meter. The recovery of D2 is not very important, because its forward current, before switchover, is only the leakage current of the device under test and therefore its stored charge is very small. However, the forward-transient overshoot voltage should also be small. The stored charge is $(I_2 - I_1)/f C$ where I_1—the reading of the meter when no forward current flows through the device under test (this current depends on the charging current of the device capacitance and its d.c. leakage current), I_2 the reading of the meter when the specified forward current flows (which is the leakage current of D2) and f the recurrence frequency of the pulse generator.

Details

The following details should be specified:
1 Test current.
2 Pulse conditions.

Saturation current test

This test measures the saturation current.

Procedure

The supply voltage is adjusted until the specified reverse voltage across the diode is achieved. The saturation current is then read from the current

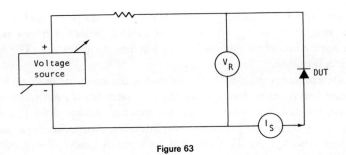

Figure 63

meter. Unless otherwise stated the reverse voltage for measurement of saturation should be approximately 80% of the nominal breakdown voltage for voltage regulators and approximately 80% of the minimum breakdown voltage for rectifiers.

Thermal resistance for signal diodes, rectifier diodes and controlled rectifiers

The purpose of this test is to measure the thermal resistance of the device.

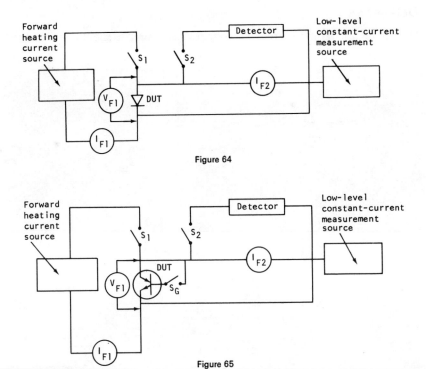

Figure 64

Figure 65

Procedure

S1 is closed for a much longer interval (heat) than it is opened (measurement). The measurement interval should be short compared with the thermal response time of the device being measured. The constant measurement interval is a small current (of the order of a few milliamperes) and so selected that the magnitude of V_{F1} changes appropriately with the device material (silicon approximately 2 mv/degC) and junction temperature. The heating source is adjustable.

The measurement is made in the following manner; the case, ambient or other reference point is elevated to a high temperature T_2 not exceeding the maximum junction temperature and the forward voltage drop V_{F1} is to be read at the start of the measurement intervals. An oscilloscope makes a convenient detector. At T_2 there will be a small power dissipated in the device due to the measurement current source.

The reference is then reduced to a lower temperature T_1 and power P_1 is applied to heat the device by increasing the current from the constant current source until the same value of V_{F1} is read as was read above. However, if P_1 is calculated as the heating power contributed by the heating current source only the equation:

$$V_{F1} = (T_2 - T_1)/P_1$$

where $P_1 = V_{F1}I_{F1}$ is sufficiently accurate.

Details

Details of test temperature, test voltages and current should be specified.

Chapter 8 Microwave diode tests

Microwave diodes

Measurement of conversion loss, output-noise ratio and other microwave parameters should be conducted with the device fitted in the holder. In the test equipment, the impedance presented to the mixer by the local oscillator and the signal generator (if used) should be the characteristic impedance of the transmission line between the local oscillator and mixer (the maximum VSWR, looking toward the local oscillator, should be 1·05 at the signal and image frequencies).

For qualification of reversible UHF and microwave devices, the radio-frequency measurements, excluding the post-environment-test end points and high-temperature-life (non-operating) end points, are to be made first, with the adapter on one end of the device and then repeated with the adapter at the opposite end of the device; for the environmental and life tests, one-half of each sample should be tested with the adapter on one end of the device and the remaining half of the sample tested with the adapter on the opposite end of the device. End-point measurements are to be made without moving the adapter.

Conversion loss test

This test determines the ratio of the available IF output power to the available RF input power.

Incremental method (Figs 66 and 67)

An expression for conversion loss is shown in the equation:

$$L_c = \frac{G_b}{2P_0(\Delta I^2/\Delta P)}$$

$$= \frac{4G_b\,(\Delta I/\Delta V)}{G_b + (\Delta I^2/\Delta V)}$$

where L_c is the conversion loss, ΔI the incremental change in current, ΔP the incremental change in power, P_0 (the average power) equals $P + \tfrac{1}{2}\Delta P$ and $G_g = 1/Z_m$. Then, since $\Delta I/\Delta V$ is the IF conductance of crystal under test,

$$\text{IF conductance} = 1/Z_{IF}$$

The crystal is loaded by the resistance $R_L + r_2$ which is adjusted to the specified load impedance Z_m. R_L is the d.c. load resistance. The current supplied by the battery balances out the diode current at some standard power level

P and makes the current in the microammeter zero. With a change in power ΔP, ΔI can be measured directly. With the injection of a small voltage (a few millivolts) ΔV at P_0 power level, ΔI can be directly measured. These values can be inserted in the equation and the conversion loss can be calculated for conditions of test.

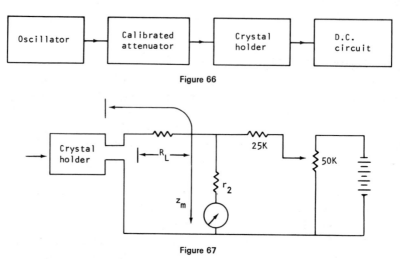

Figure 66

Figure 67

Heterodyne method

A noise generaor feeds signal power to the mixer which converts the power to the 'intermediate' frequency by beating with the local oscillator. The converted power is measured with an IF wattmeter. Both the available noise power from the generator at A1 shown on Fig. 68 and the increase in the available IF power at B should be measured when the noise is applied, their ratio being the conversion loss.

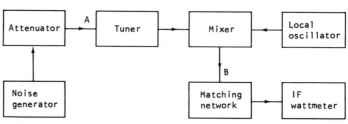

Figure 68

Modulation method

Conversion loss is given by the equation:

$$L_c = \frac{4n}{(1+n)^2} \cdot \frac{m^2 p}{G_b E_B{}^2}$$

where M is the modulation coefficient, P the available power, E_B the r.m.s. modulation across load and n the ratio of load conductance IF.

To avoid measuring G_b for each unit the factor $4n/(1+n)^2$ is assumed to be unity. The error caused by this approximation is less than 0·5 dB and is in such a direction to make a unit with an extreme conductance seem worse:

$$L_c = m^2 p / G_b E_B^2$$

Since the modulation coefficient is difficult to measure, this equipment is calibrated with standard

$$L_c \text{ (dB)} = 10 \log_{10} (m^2 P / G_b)\ 20 \log_{10} E_B^2$$

A Ballantine-type voltmeter can be used to measure $20\log_{10} E_B^2$ directly. The voltmeter is set on the 0·01 V full-scale and the modulation voltage set so that the term $10\log_{10}(m^2/G_b)$ is equal to 20 on the dB scale.

To obtain this setting, the modulation is adjusted, so that the voltmeter reading on the decibel scale is 20·0 minus the value of conversion loss for the standard crystals. This corresponds to a value of M of 1·58% for $P = 1·0$ mW and $G_b = 1/400\ \Omega$. The conversion loss for unknown crystals is then 20·0 minus the reading of the output meter in decibels.

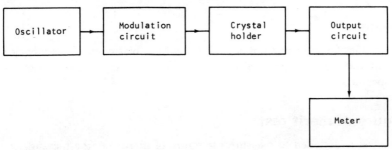

Figure 69

Over-all noise figure method

The over-all noise figure method derives the conversion loss by known properties of the apparatus and is expressed by the equation:

$$N_F = L_c (N_{IF} + N_r - 1)$$

where L_c is the conversion loss (power ratio), N_r the noise ratio (power ratio) and N_{IF} the IF amplifier noise figure (power ratio).

With the measured noise temperature, N_r, conversion loss can be calculated. The load impedance, test conditions and lead resistance are to be specified.

Detector power efficiency test

The purpose of this test is to measure the detector power efficiency.

Procedure

Resistor R_L and capacitor C_1 (Fig. 70) comprise the load circuit and should be as specified. R_1, in conjunction with RL_1 provides the bias current for the device under test. Capacitor C_2 provides RF bypass for the output current

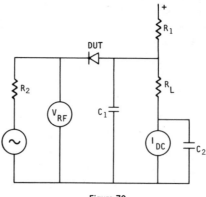

Figure 70

meter I_{DC}. The frequency and amplitude of the a.c. signal and the output current $I_{a.c.}$ is measured when the a.c. signal is applied. Then

$$\text{Detector power efficiency} = [4(\Delta I_{ac})^2 R_L R_G / V^2 (\text{r.m.s.})]100\%$$

Details

The values for current components R_L and C_1, bias current, frequency and amplitude of the a.c. signal also the impedance of the signal generator are to be specified.

Figure of merit test

This test is designed to measure the figure of merit, i.e. the measure of the excellence of a video crystal in a video receiver. This is the case since all video amplifiers have essentially the same equivalent noise-generating resistance, R_A, in series with the grid circuit in the input stage. With the absence of bias, the noise generated by a crystal in the microwatt range of RF power is almost entirely the Johnson noise of a resistance equal to the impedance of the crystal. The 'figure of merit' is represented by:

$$M = S_i Z_V / (Z_V + R_A)^{\frac{1}{2}}$$

where M is the figure of merit, S_i the short-circuit current sensitivity ($\mu A/\mu W$), Z_V the video impedance and R_A the equivalent noise-generating resistance (1200 Ω).

An expression relating rectified current (i) to figure of merit (M), power (P) and load resistance (R_x) is

$$i = (PM/2\sqrt{R_x})(1 + \Delta)^{\frac{1}{2}}/(1 + \tfrac{1}{2}\Delta)$$

R_x is dependent upon the upper and lower limits of Z_V is calculated from

$$R_x = [(R_1 + 1200)(R_2 + 1200)]^{\frac{1}{2}}$$

where R_1 is the lower limit and R_2 the upper limit.

For devices falling within the range R_1 to R_2 the correction factor $(1 + \Delta)^{\frac{1}{2}}/(1 + \frac{1}{2}\Delta)$ is close to unity, and a good approximation is

$$i = PM/2\sqrt{R_x}$$

Then M may be determined by measuring i under proper conditions.

Power, P, is adjusted as specified and current i, is measured (see Fig. 71).

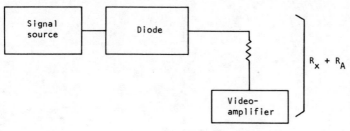

Figure 71

Intermediate frequency impedance test

This test measures the impedance at the IF output terminals of the mixer, under specified operating conditions. Small-signal conditions should prevail during this measurement.

Test circuit

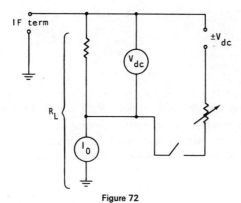

Figure 72

Procedure

At intermediate frequencies to 65 MHz, this impedance will be equal to the small signal d.c. resistance at the operating point, and any of the following methods may be employed to measure IF impedance.

D.c. method
A small d.c. voltage (0.005 V max) is injected in series with the device (see Fig. 72) and the change in current is noted with:

$$Z_{if} = dV/dL$$

A.c. method

A low level (50 μA) constant current audio-frequency is injected in series with the device and the audio frequency voltage appearing across the diode is measured. This measurement technique can be conveniently effected with the circuit shown in Fig. 73.

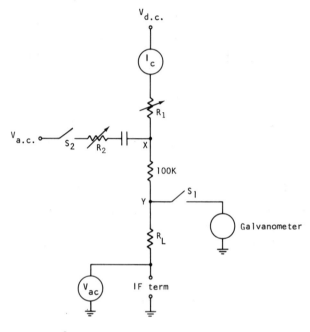

Figure 73

The d.c. voltage is used to oppose the voltage developed across R_L by rectified device current. With S1 closed, R_1 is adjusted for a null reading on the galvonometer establishing a virtual ground at point Y. S1 may then be opened and while the device d.c. load requirement is satisfied, no shunt impedance remains across the device at the d.c. test frequency.

A 50 μA constant current level is established by adjusting R_2 for 5V a.c. at point x when S1 and the IF terminals are open-circuited. Caution should be exercised when inserting or removing devices from test holders. To maintain a low d.c. voltage at point Y and prevent device damage from high surge currents, S1 should be closed during these operations.

Impedance-bridge method

An impedance bridge operating at any signal frequency up to 65 MHz is connected to the output terminals of the mixer and the IF impedance is read directly from the bridge (see Fig. 74). The bridge-signal level should be below that at which a 1·0% increase in rectified device current occurs when the signal is applied.

Microwave diode tests 101

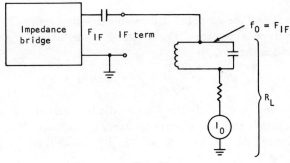

Figure 74

Details

Test conditions should be specified.

Output noise ratio test

This test measures the output noise ratio which is the ratio of the available 30 MHz noise power of the device when excited by a local oscillator under the specified test conditions to that of a resistor at a standard temperature, 293 ± 5 K.

The measurement should be made in a circuit equivalent to that shown in Fig. 75.

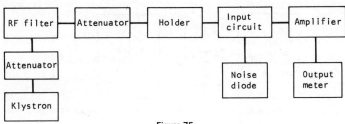

Figure 75

The RF filter has a Q as specified under standard test conditions. It is tuned to the standard test frequency. The filter is to reject the noise sidebands of the local oscillator tube. The input coupling unit is a non-dissipative network designed to make the noise ratio measurement independent of the device within the Z_{IF} of the type. The tuning of the coupling circuit is accomplished by using resistors in the holder (covering the range of IF impedance of the type) as a signal generator, made 'noisy' by current from a noise diode.

Equivalent noise ratio should result in identical indicated noise ratio independent of actual value of IF impedance. The output meter utilises a crystal rectifier as a detector giving indications nearly proportional to power.

Calibration

The input circuit of the amplifier is terminated with a resistor falling within the device IF impedance range. The resistor is made 'noisy' by a noise diode. The noise ratio is then

$$NR_0 = \frac{eIR}{2kT_0} + 1 = 20IR + 1$$

where $T_0 = 293 \pm 5$ K, I the noise diode current in amperes, k Boltzman's constant and e the electronic charge ($1 \cdot 60 \times 10^{-19}$C). The output meter is calibrated directly in terms of NR_0 by this method.

Procedure

The ratio of the noise output power from the device under test to that from the standard resistor, at room temperature, is the noise ratio and is read directly from the output meter.

Details

The test conditions should be specified.

Video impedance test

This test measures the video impedance of the device. The test should be conducted in the detector and at the frequency specified. Small-signal conditions must prevail during this measurement.

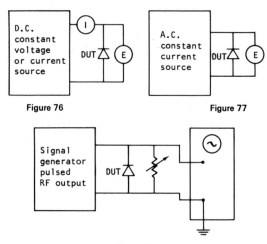

Figure 76

Figure 77

Figure 78

Calibration and procedure (Figs 76–78)

D.c. method

A small d.c. signal (less than 5 mV) is applied to the device through a constant voltage or current source. Current or voltage is measured with a low resistance (less than 5 Ω) microammeter, or high-resistance voltmeter, respectively.

$$Z_V = E/I$$

A.c. method

A small a.c. current (less than 0·1 μA) is passed through the device from a constant current a.c. source. The a.c. voltage is measured across the device with a millivoltmeter.

$$Z_V = E/I$$

Pulsed RF method

A pulsed RF as specified is fed to the device whose output is fed into the vertical amplifier of an oscilloscope. A resistor is placed in parallel with the device and varied to lower the rectified pulse to half its value. Z_V equals the resistance required to halve the pulse.

Burnout by repetitive pulsing

This test determines the capabilities of the device to withstand repetitive pulses.

Procedure

The device should be subjected to a pulse or pulses of the width, voltages and repetition rate specified. The pulse polarity should be such as to cause the current to flow in the forward direction. Burnout should not be considered as occurring when there is a catastrophic failure or when the maximum charge in the specified electrical parameter is exceeded.

Details

The following details are to be specified:

1. Pulse width.
2. Pulse voltage.
3. Repetition rate.
4. Pulse-generator internal resistance.

Burnout by single pulse

The purpose of this test is to determine the capability of the device to withstand a single voltage pulse.

Procedure

The device should be subjected to a pulse from the coaxial line specified. The line should be charged with the specified voltage and contact made by dropping the centre conductor vertically from a height of 5 cm above the contact position. The electrical and mechanical connection should be such as to have minimum effect of the free fall of the conductor. The polarity of the inner conductor with respect to the outer conductor should be such as to cause the device current to flow in the forward direction.

Details

The test voltage and polarity are to be specified.

Chapter 9 Thyristors

Holding current test

This test measures the holding current of the devices.

Procedure

The value of resistance in the anode circuit R_1 of Fig. 79 is fixed and the anode supply voltage is adjusted so that the forward current that will flow when the device is turned on is high enough to assure that the device is completely fixed. The anode voltage is then decreased gradually thus decreasing the forward current until the device turns off.

The value of the forward current immediately prior to the device turning off is the holding current.

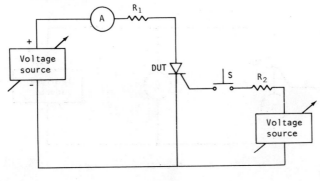

Figure 79

Details

The following details are to be specified:
1. Anode voltage.
2. Gate voltage.
3. Resistances R_1 and R_2

Forward leakage current test

This test measures the forward leakage current of the device.

Procedure

The supply voltage is adjusted to obtain the specified value of forward voltage across the device Fig. 80. The forward leakage current is then read from the current meter.

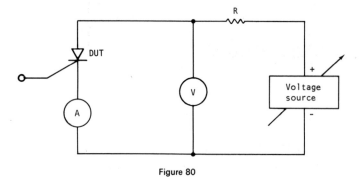

Figure 80

Details

Test voltage should be specified.

Reverse leakage current test

Procedure

The supply voltage (see Fig. 81) is adjusted to obtain the specified value of reverse across the device. The reverse leakage current is then read from the current meter.

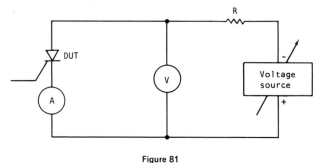

Figure 81

Details

The test voltage is to be specified.

Pulse response test

Procedure

The pulse response of the device should be measured in the circuit of Fig. 82. R_2 is adjusted to permit the specified value of forward current to flow in the device being measured. When it is the 'on' state, C, R_1 and the secured controlled rectifier D2 are used to switch off the device being measured. C should be large enough to ensure that the device will turn off. R_1 limits the recurrent peak reverse current to below the rated value. The pulse repetition rate should be low enough to ensure that the anode-cathode voltage of the device being measured reaches the value of forward working voltage specified for the measurement.

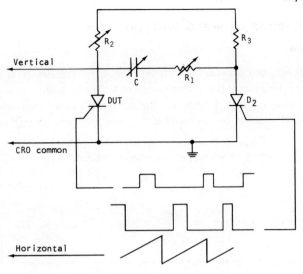

Figure 82

Details

The following details should be specified:

1. Anode voltage.
2. Resistance R_2.
3. Test current.
4. Repetition rate.

Gate triggering signal test

This test measures the gate triggering signal.

Procedure

The gate voltage is slowly increased from zero (see Fig. 83). The gate current and/or gate voltage are read immediately prior to the drop in forward voltage. The anode voltage and resistances R_1 and R_2 should be specified.

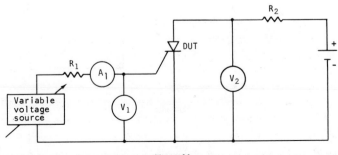

Figure 83

Instantaneous forward voltage drop

Procedure

A specified forward anode d.c. current is set after the device is turned 'on', (see Fig. 84). Following the achievement of thermal equilibrium, the forward voltage drop is measured between the terminal of the controlled rectifier. The source of d.c. for the forward voltage drop characteristic testing is not

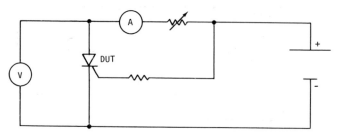

Figure 84

considered important provided the ripple is less than 5%. The voltmeter should be connected directly to the anode and cathode terminals of the device under test so that the voltmeter will read the actual value of forward current drop and will not be affected by voltage drops across the terminals or leads. Anode current and time before measurement should be specified.

Rate of voltage rise test

The purpose of this test is to ensure that the device does not switch on under the specified conditions.

Procedure

A specified value of forward voltage should be applied to the device as shown in Fig. 85.

The device should remain in the blocking state during the course of the test.

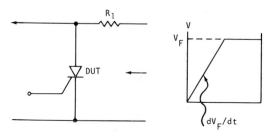

Figure 85

Details

The forward anode voltage and the rate of voltage rise should be specified.

Chapter 10 Tunnel diode tests

Junction capacitance
This test determines the small-signal junction capacitance of the tunnel diode.

Procedure
Since junction capacitance is a function of bias, it is necessary to specify the forward bias C_1 is to be determined. The true value of junction capacitance, at a given bias, is obtained by subtracting the capacitance of the diode package from the observed capacitance. Isolation of the d.c. power supply from the complex impedance bridge (see Fig. 86) is effected by the R_1, L, C_2 branch of the circuit.

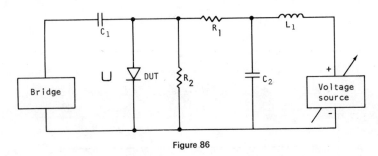

Figure 86

Details
The following should be specified:
1 Values for the circuit elements R_1, C_1, C_2, L_1 and R_2.

Static characteristics of tunnel diodes test
This test measures the static characteristics V_p, V_V, I_V, V_{FP} and R_d of the tunnel diode.

Procedure
For the measurement of the static characteristics by the point-by-point method, the circuit of Fig. 87 should be used. Resistance R_2 is small to obtain low voltage and low impedance. Resistor R_3 is a current measuring resistor. Resistance R_1 is much larger than R_2. To obtain a plot in the negative resistance region, R_1 should be less than the magnitude of the incremental negative resistance of the tunnel diode.

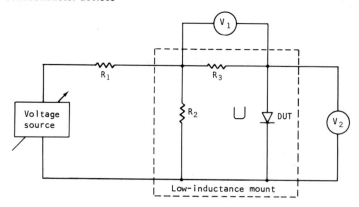

Figure 87

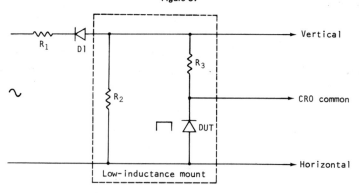

Figure 88

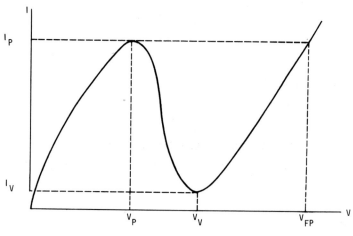

Figure 89

For the measurement of the static forward characteristics of the device by oscillographic means the circuit shown in Fig. 88 is to be used. The magnitude of R_1 should be less than the magnitude of the incremental negative resistance of the tunnel diode. Resistor R_3 is a current measuring resistor and its resistance should be chosen to give a suitable CRO deflection. Since the negative resistance is represented by the inverse slope of the I–V curve (see Fig. 89) between peak and valley voltage points, its approximate value can be estimated from the curve. Resistances R_1, R_2 and R_3 and the signal frequency should be specified.

Series inductance test

This test measures the value of the small signal series inductance of the device.

Procedure

The device should be reverse-biased for the series inductance measurement.

A sufficiently high frequency signal should be employed to emphasis the inductive reactance, but not high enough to allow capacitive parasitics to short circuit the device, thus precluding the determination of L_S. A recommended frequency device is one approximately one-fourth of the self resonant frequency of the device under test.

Isolation of the d.c. power supply from the complex impedance bridge is accomplished by the choke L_1, in conjunction with C_1, R_1, C_2 branch (Fig. 90).

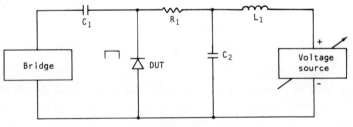

Figure 90

Details

The following details should be specified:

1 Value for circuit elements, R_1, L_1, C_1 and C_2.
2 Signal frequency.
3 Reverse bias at which L_S is measured.

Negative resistance test

The purpose of this test is to determine the magnitude of the negative resistance.

Procedure

The resistance R_1 should be less than the incremental negative resistance of the tunnel diode. Resistor R_3 is a current limiting resistor and its resistance should be chosen to give a suitable CRO deflection. Diode D1 is a half-wave rectifier.

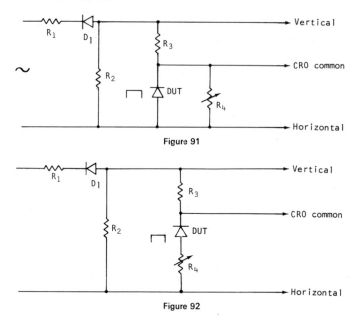

Figure 91

Figure 92

Short-circuit stable method
Shunt the tunnel diode with a variable resistor R_4 (Fig. 91). Vary R_4 until the slope of the negative resistance appears horizontal (zero slope) on the curve trace. The shunting resistance is now equal to the magnitude of R_d the negative resistance ($R_4 = R_d$).

Open-circuit stable method
In series with the tunnel diode connect a variable resistor R_4, (Fig. 92). Vary R_4 until the slope in the negative region appears vertical (infinite slope) on the curve trace. The series resistance R_4 is now equal to the magnitude of the negative resistance ($R_4 = R_d$).

Details
The following details are to be specified:
1 Source impedance, R_1.
2 Current sensing resistor, R_3.
3 Variable resistor, R_4.

Series resistance test
This test determines the series resistance of the device.

Procedure
The measurement of the series resistance should be accomplished for the device when biased in the reversed direction (see Fig. 93). The linearity of the ohmic region should be assured and the value of the power dissipation should be such that no error is introduced as a result of excessive diode heating.

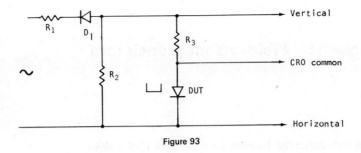

Figure 93

The slope of the linear portion of the reverse biased tunnel diode should be sealed within a specified accuracy in the direct determination of the series resistance of the device.

Details
The following details are to be specified:
1. Current-sensing resistor R_3.
2. Reverse bias at which R_3 is measured.

Switching time test
This test measures the switching time of the tunnel diode.

Procedure
A block diagram of the measuring circuit is shown in Fig. 94.

To perform the switching time measurement, it is necessary that the maximum generator current be greater than the diode peak current, and that the changes in generator current during the measurement time be negligible compared with I_b. The oscilloscope input impedance to be such that the current absorbed by the probe is at all times less than the peak current of the diode.

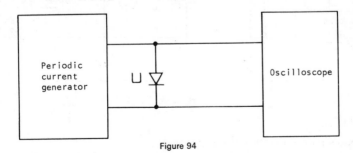

Figure 94

Details
The following details should be specified:
1. Generator current.
2. Repetition rate.
3. Rise time of oscilloscope.

Chapter 11 Field-effect transistors

Gate-to-source breakdown voltage test

This test determines if the breakdown voltage under the specified conditions is greater than the minimum limit.

Procedure

The ammeter should present essentially a short-circuit to the terminals between which the current is being measured or the voltmeter readings should be corrected for the drop across the ammeter (see Fig. 95).

The resistor R_1 is a current-limiting resistor and should be of sufficiently high resistance to avoid excessive current flowing through the device and current meter. The voltage should be gradually increased, with the specified bias condition A, B, C or D applied, from zero until either the minimum limit for $V_{(BR)GSX}$ or the specified test current. The device is acceptable if the minimum limit for $V_{(BR)GSX}$ is reached before the test current reaches the specified value.

If the specified test current is reached first, the device should be considered a failure.

$V_{(BR)GSX}$ is the breakdown voltage, gate to source, with the specified bias condition applied from drain to source.

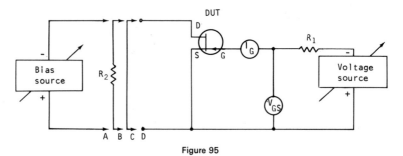

Figure 95

Details

The following details should be specified:
1 Test current.
2 Bias condition:
    A—Drain to source: reverse bias.
    B—Drain to source: resistance return.
    C—Drain to source: short-circuit.
    D—Drain to source: open-circuit.

Gate-to-source voltage or current test

This test measures the gate-to-source voltage or current of the field-effect device.

Procedure

The ammeter should present a short-circuit to the terminals between which the current is being measured or the voltmeter readings corrected for the drop across the ammeter (see Fig. 96). The voltage source and bias source to be adjusted to bring V_{GS} and I_D to their specified values. The voltage V_{GS} or current I_G may then be read.

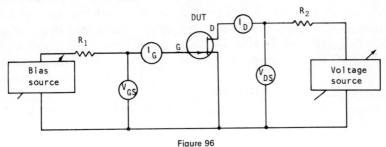

Figure 96

Details

The test voltages, currents and parameter to be measured should be specified.

Drain-to-source 'on'-state voltage test

This test measures the drain to source voltage of the field-effect transistor at the specified value of drain current.

Procedure

The specified bias condition is to be applied between the gate and source and the voltage source should be adjusted to bring I_D to the specified value. The voltage V_{DS} may then be read (see Fig. 97).

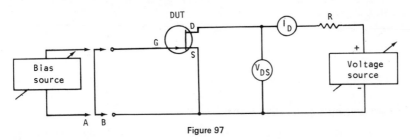

Figure 97

Details

The following details should be specified:
1. Test current.
2. Gate-to-source bias condition:
   A. Voltage-biased (voltage and polarity).
   B. Short-circuited.

Drain-to-source breakdown voltage test

This test determines if the breakdown voltage of the field-effect device under specified conditions is greater than the specified minimum limits.

Procedure

The ammeter is to present a short circuit to the terminals between which the current is being measured or the voltmeter readings corrected for the drop across the ammeter (see Fig. 98).

The resistor R_1 is a current-limiting resistor and should be of a high enough resistance to avoid excessive current flowing through the device and current meter. The voltage is to be gradually increased from zero, with the specified bias condition A, B, C or D applied, until either the minimum limit for $V_{(BR)DSX}$ or the specified test current is reached. The device is acceptable if the minimum limit for $V_{(BR)DSX}$ is reached before the test current attains the specified value, if the specified test current is reached first, the device should be considered a failure.

$V_{(BR)DSX}$ is the breakdown voltage, drain to source, with specified bias condition applied from gate to source.

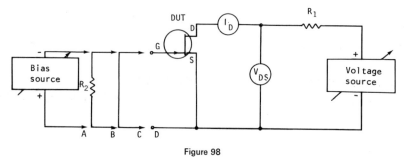

Figure 98

Details

The following details should be specified:
1. Test current.
2. Bias condition:
    A—Gate to source: reverse bias.
    B—Gate to source: resistance return.
    C—Gate to source: short-circuit.
    D—Gate to source: open circuit.

Gate reverse current test

This test measures the gate reverse current of the field-effect device under specified conditions.

Procedure

The ammeter should present a short-circuit to the terminals between which the current is being measured or the voltmeter readings should be corrected for the drop across the voltmeter (see Fig. 99).

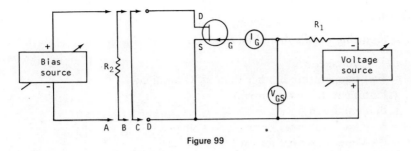

Figure 99

The specified d.c. voltage should be applied between the gate and the source with the specified bias conditions A, B, C or D, applied to the drain. The measurement of current is to be made at the specified ambient or case temperature.

Details

The following details should be specified:

1. Test voltage.
2. Test temperature if other than 25°C ambient.
3. Bias condition:
   A—Drain to source: reverse bias and voltage.
   B—Drain to source: resistance R_2.
   C—Drain to source: short-circuit.
   D—Drain to source: open-circuit.

Drain current test

The purpose of this test is to measure the drain current of an FET.

Procedure

The ammeter in the case should present a short-circuit to the terminals between which the current is being measured or the voltmeter readings are to be corrected for the drop across the ammeter (see Fig. 100).

The specified voltage is to be applied between the drain and source with the bias condition A, B, C or D applied to the gate. The measurement of current should be made at the specified ambient or case temperature.

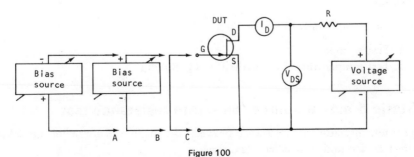

Figure 100

Details

The following details should be specified:

1. Test voltage.
2. Test temperature if other than 25°C ambient.
3. Parameter to be measured.
4. Bias condition:
    A—Gate to source: reverse bias voltage.
    B—Gate to source: forward bias voltage.
    C—Gate to source: short-circuit.
    D—Gate to source: open-circuit.

Drain reverse current test

This test measures the drain reverse-current of the field-effect device under specified conditions.

Procedure

Again the ammeter is to present a short-circuit to the terminals between which the current is being measured or the voltmeter readings should be corrected for the drop across the ammeter (see Fig. 101).

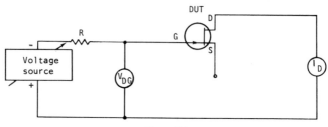

Figure 101

The specified d.c. is to be applied between the drain and the gate. The measurement of current to be made at the specified ambient or case temperature.

Details

The following details to be specified:

1. Test voltage.
2. Test temperature if other than 25°C ambient.

Static drain-to-source 'on'-state resistance test

This test measures the resistance between the drain and source of the field-effect device under specified conditions.

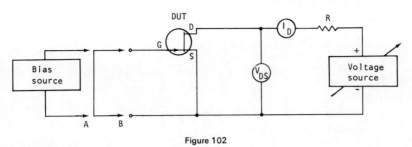

Figure 102

Procedure

The specified bias is to be applied between the gate and source and the voltage source to be adjusted so that the specified current is achieved. The drain to source voltage is then measured (see Fig. 102). Then

$$r_{DS(on)} = V_{DS}/I_D$$

Details

The following details should be specified:

1 Test currents.
2 Gate-to-source bias condition:
   A—Voltage-biased, voltage and polarity.
   B—Short-circuited.

Small-signal, drain-to-source 'on'-state resistance test

This test measures the resistance between the drain and source of the field-effect device under specified conditions.

Procedure

The a.c. voltmeter is to have an input impedance high enough that halving it does not change the measured value within the required accuracy of the measurement (see Fig. 103).

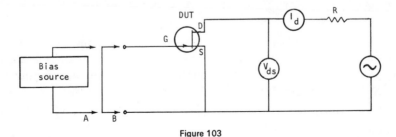

Figure 103

The specified bias condition is to be applied between the gate and the source and an a.c. sinusoidal signal current, I_d of the specified r.m.s. value applied. Then

$$r_{ds(on)} = V_{ds}/I_d$$

Details

The following details to be specified:

1. Test current.
2. Test frequency.
3. Gate-to-source bias condition:
   A—Voltage-biased, bias voltage and polarity.
   B—Short-circuited.

Small-signal, common source, short-circuited input capacitance test

This test measures the input capacitance of the field-effect device under specified small-signal conditions.

Procedure

The circuit and procedure shown (Fig. 104) are for common-source configuration. For other configurations the circuit and procedure should be changed accordingly.

The capacitors C_1 and C_2 should present short-circuits at the test frequency

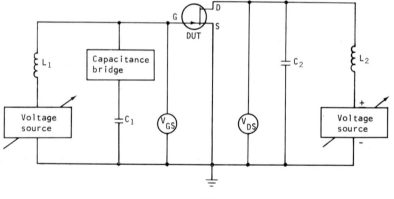

Figure 104

—L_1 and L_2 to present a high a.c. impedence at the test frequency for isolation. The bridge to have low d.c. resistance between its output terminals and should be capable of carrying the test current without affecting the desired accuracy of measurement.

Details

The following details should be specified:

1. Test voltages and current.
2. Measurement frequency.
3. Parameter to be measured.

Field-effect transistors   121

Small-signal, common source, short-circuit, reverse-transfer capacitance test

This test is to measure the reverse-transfer capacitance of the field-effect device.

Procedure

The circuit and procedure shown (Fig. 105) are for common-source configuration. For other configurations the circuit and procedure should be changed accordingly. Terminal 2 of the bridge is to be the terminal with an a.c. potential closest to the a.c. potential of the guard terminal so as to provide an effective short circuit of the input.

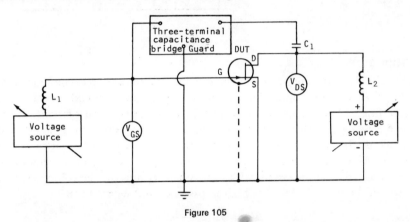

Figure 105

The dotted connection between the case and ground should be used for devices that are not internally electrically connected to any element. If the case is internally connected to any element, the dotted connection must not be used. The capacitor C_1 should present a short-circuit at the test frequency and L_1 and L_2 present a high a.c. impedance for isolation. The bridge should have low d.c. resistance between its output and should be capable of carrying the test current without affecting the desired accuracy of measurement.

Details

The following details should be specified:

1 Test voltages and currents.
2 Measurement frequency.
3 Parameter to be measured.

Small-signal, common-source, short circuit, output admittance test

This test measures the output admittance of a field-effect device, under specified small-signal conditions.

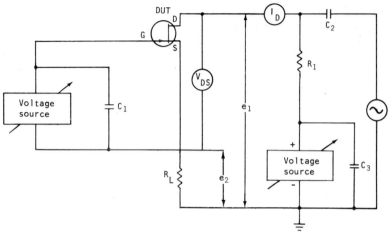

Figure 106

Procedure

The circuit and procedure shown are for common-source configuration. For other configurations the circuit and procedure should be changed accordingly. The capacitors C_1, C_2 and C_3 are to present short circuits at the test frequency in order to effectively couple and by-pass the test signal. R_1 and R_L should be short circuits compared with the output impedance of the device. After setting the specified d.c. conditions. the V_{DS} meter should be disconnected from the circuit whilst measuring e_1 and e_2. The voltages e_1 and e_2 should be measured with a high impedance voltmeter. Then

$$Y_{os} = I_d/(e_1 - e_2)$$

where

$$I_d = e_2/R_L$$

Thus

$$Y_{os} = e_2(e_1 - e_2)/R_L$$

Details

The following details should be specified:
1. Test frequency.
2. Test voltages and currents.
3. Parameter to be measured.

Small-signal, common-source short circuit, forward transadmittance test

This test measures the forward transadmittance of the field-effect device under specified small-signal conditions.

Procedure

The circuit and procedure shown (Fig. 107) are for common-source configuration. For other configurations the circuit and procedure should be changed.

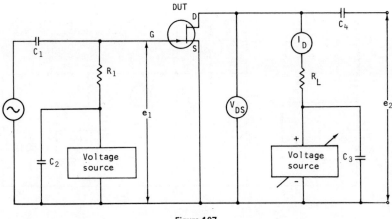

Figure 107

The capacitors C_1, C_2, C_3 and C_4 are to present short-circuits at the test frequency in order to effectively couple and by-pass the test signal.

R_1 should be a short-circuit compared with the input impedance of the device. R_L should be a short-circuit compared with the output impedance of the device. The voltages e_1 and e_2 are to be measured with high-impedance a.c. voltmeters. Then

$$Y_{fs} = I_d/e_1$$

where

$$I_d = e_2/R_L$$

Thus

$$Y_{fs} = \frac{e_2}{e_1} \cdot \frac{1}{R_L}$$

Details

The following details are to be specified:

1. Test frequency.
2. Test voltages and currents.
3. Parameter to be measured.

Small-signal, common source, short-circuit, input admittance test

The purpose of this test is to measure the input admittance of the field-effect device under the specified small-signal condition.

Procedure

The circuit and procedure shown (Fig. 108) are for common-source configuration. For other configurations the circuit and procedure should be changed accordingly.

The capacitors C_1, C_2 C_3 and C_4 should present short-circuits at the test frequency in order to effectively couple and by-pass the test signal. R_1 facilitates the adjustment of e_1. Its use is optional. R_2 should be such that d.c.

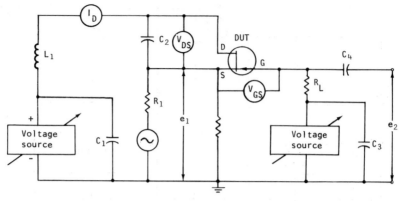

Figure 108

biasing is possible. R_L should be a short circuit compared with the input impedance of the device. V_{DS} should be adjusted to the specified value, then the gate voltage supply is to be adjusted so that V_{GS} or I_D equals the specified value, and the voltage e_1 and e_2 should be measured. Then

$$Y_{is} = I_g/(e_1 - e_2)$$

where

$$I_g = e_2/R_L$$

Thus:

$$Y_{is} = e_2/R_L(e_1 - e_2) \quad \text{or} \quad Y_{is} = e_2/R_L \, (e_1 - e_2)$$

e_1 must be greater than e_2; therefore

$$Y_{is} = e_2/R_L \, e_1$$

Details

The following details are to be specified:
1. Test frequency.
2. Test voltages and currents.
3. Parameter to be measured.

Note

Circuits are shown in this chapter for N-channel field-effects devices in one circuit configuration. They may readily be adapted for P-channel devices and for other configuration.

Bibliography

KLASENS, H. A. and KOELMANS, H. *Solid-state Electronics*, **7,** 701–702 (1964).

DAVEY, J. E. and PANKEY, H., *A. appl. Phys.*, **35,** 2203–2209 (1964).

SEVIN, L. J., 'Effect of temperature on FET characteristic, *Electro-Technology*. **73,** 103–707 (April 1964). (Effects of temperature on the drift mobility and 'built-in' voltage of FETS and how they interact to permit temperature compensation).

TATOM, C., and WILCOX, D., 'How to measure Y-parameters of field-effect transistor's, *Electronic Equipment Engineering*, **11,** 95–97 (November 1963). (Technique for measuring low-frequency y-parameters of FETs at automatically-fixed drain-source currents and voltages).

WARNER, R. M., 'Expitaxial FET cut-off voltage', *Proc. IEEE*, **51,** 939–940 (June 1963). (Correspondence regarding a study of V_p on resistance to verify an assumed parabolic impurity distribution.)

ZIEL VAN DER., A., 'Gate noise in field-effect transistors at moderately high frequencies', *Proc. IEEE*, **51,** 461–467 (March 1963). (Gate noise in the FET and its cause due to a capacitive coupling between the channel and gate.)

COWLES, L. G., 'The definition and measurement of FET pinch-off voltage', *Proc. IEEE*, **52,** 200 (February 1964).

EVANS, A. D., 'Measurement of I_{GSS} in field-effect transistors', *Electrical Design News*, **9,** 122–123 (March 1964).

PALMER, W., 'Evaluating breakdown voltage characteristicts of field-effect transistors', *Electron Equipment Engineering*, **11,** 54–57 (May 1963).

TOMPKINS, J. D., 'An inside Look at FET terminology and parameters, *Electrical Design News*, **9,** 50–52, 52–58 (July 1964).

HOLLAND, C. E., 'FET's synthesize negative resistance', *Electro-Technology*, **73,** 124–126 (April 1964). (Two FETs connected together to produce a composite device which has a relatively linear negative resistance region.)

BRUNCKE, W. C., 'Noise measurements in field-effect transistors', *Proc. IEEE*, **51,** 378–379 (February 1963). (Correspondence giving a comparison of measured results of noise in FETs with van der Ziel's noise theories and an extension of the theory to explain discrepancies found.)

CSANKY, G., and WARNER, R. M., 'Put more snap in logic circuits with field-effect transistors, *Electronics*, **36,** 43–45 (14 June 1963). (The use of a second FET as the drain load resistor to increase the effective load resistance and hence the overall voltage gain of the inverter stage.)

DACEY, G. C., and ROSS, I. M., 'The field-effect transistor', *Bell System Tech. Journal*, **34**, 1149–1189 (November 1955). (A rather thorough treatment of FET theory in which it is shown that the cut-off frequency is inversely proportional to the value of V_p. Includes design nomographs and a description of the fabrication and performance of several FETs.)

Index

A.c. testing, 9–10
Accelerated tests, 12–14
   elimination of rogue failures, 13–14
Acceleration, constant, test for, 33
Accuracy, meters, 1
Admittance:
   field-effect transistors, small-signal, common source, short-circuit:
      input, 123–4
      output, 121–2
   high- and low-frequency tests:
      forward-transfer, 68
      input, 67–8
      output, 65, 69–70
      reverse-transfer, 69
Air supply, salt-spray test, 26
Alternate sweeps, a.c. testing, 10
Altitude, reduced barometric pressure, environmental tests, 15–16
Analogue voltmeters, a.c. testing, 10
Atomisation, salt-spray test, 28
Atomisers, salt-spray test, 26
Average reverse current, general diode tests, 89–90

Barometric pressure, reduced, environmental tests, 15–16
Bending stress, mechanical tests, 40–1
'Bleed' load, 7
'Breadboards', reliability of testing, 4
Breakdown:
   impedance, general diode tests, 89–90
   voltage:
      electrical tests of transistors:
         collector-to-base, 44
         collector-to-emitter, 45–6
         emitter-to-base, 47–8
      field-effect transistors:
         drain-to-source, 116
         voltage gate-to-source test, 114
      general diode tests, 83
         temperature coefficient, 90–1
Burnout, by pulsing:
   electrical tests of transistors, 45
   microwave diode tests, 103–4

Capacitance:
   field-effect transistors:
      input capacitance test, 120
      reverse transfer, 121
   general diode tests, 81
      variable, quality factor, 87–8
   high- and low-frequency tests:
      depletion-layer, 73–4
      diffusion, 72–3
      direct inter-terminal, 72
      input, 71
      output, 70–1
   tunnel diodes, junction, 109
Case temperature, hex-base devices, 31–2
Centrifugal force, mechanical tests, 33
Chambers, salt-spray test, 26
Charge, stored, general diode tests, 91–2
Circuit performance, measurement, 57–63
   general, 62
   thermal resistance:
      collector cut-off method, 57–8
      d.c. current gain, continuous method, 62–3
      d.c. forward voltage drop, emitter-to-base continuous method, 59–60
      forward voltage drop, collector-to-base diode method, 60–1
      forward voltage drop, emitter-to-base diode, 58–9
      thermal response time test, 60
      thermal time constant test, 61–2
Coatings, salt-spray test, 25–9
Collectors:
   breakdown voltage, 44, 45–6
   current tests, 48, 49–50
   thermal resistance and circuit performance, 57–8, 60–1
   voltage tests, 49–50
Constant acceleration, test for, 33
Controlled:
   environments, 14
   rectifiers, thermal resistance tests, 93–4
Conversion loss, microwave diode tests, 95–7

128 Semiconductor devices

Corrosion:
  salt-atmosphere (corrosion) test, 24–5
  salt-spray test, 25–9
Crystals, video, figure of merit test, 98–9
Current:
  forward *See* Forward: current
  forward leakage *See* Forward: leakage current
  gain, d.c. measurement of thermal resistance and circuit performance, 62–3
  holding *See* Holding current
  leakage, environmental tests, 17–19
    See also Forward: leakage current
  output *See* Output: current
  reverse *See* Reverse: current
  reverse leakage *See* Reverse: leakage current
  saturation *See* Saturation: current
  surge *See* Surge: current
  tests:
    field-effect transistors:
      drain, 117–18
      gate-to-source, 115
    transistors:
      collector-to-base, 48
      collector-to-emitter, 49–50
      emitter-to-base, 50–1
Cut-off:
  current, transistors, 48, 49–50, 50–1
  forward-current transfer ratio, 78–9
  measurement of thermal resistance and circuit performance, 57–8
Cycles, number of, moisture resistance tests, 20–1

D.c. testing, 7–9
Depletion-layer capacitance, 73–4
Design:
  for reliability, 3–4, 10–13
    quantitative reliability, 13
    step-stress testing, 12–13
Detector power efficiency, microwave diode tests, 97–8
Dew point, environmental tests, 32
Diffusion capacitance, high- and low-frequency tests, 72–3
Digital voltmeters:
  a.c. testing, 10
  loading effects, 6
Diodes:
  general tests, 81–94
    average reverse current, 89
    breakdown voltage, 83
      temperature coefficient, 90–1

capacitance, 81
d.c. output current, 81–2
forward current voltage, 82
forward recovery time, 83–4
quality factor for variable capacitance diodes, 87–8
rectification efficiency, 88–9
reverse current and reverse voltage, 82–3
reverse recovery time, 84–7
saturation current, 92–3
small-signal breakdown impedance, 89–90
small-signal forward impedance, 91
stored charge, 91–2
surge current, 90
terminal strength, 38–9
thermal resistance, 93–4
  microwave *See* Microwave diodes
  tunnel *See* Tunnel diodes
  UHF, terminal strength test, 38–9
Dipping device, solderability tests, 35
Direct inter-terminal capacitance test, 72
Discontinuity, dew point tests, 32
Drain, field-effect transistors:
  current test, 117–18
  reverse current test, 118
  static drain to source, 'on'-state resistance test, 118–19
  to-source breakdown voltage, 116
  to-source, 'on'-state resistance test, 119–20
  to-source 'on'-state voltage test, 115
Drift, electrical tests of transistors, 46
Dross, solderability, 37

Electrical:
  measurements, test conditions, 3
  noise *See* Noise
  tests:
    frequency, 1
    semiconductor (transistor) *See* Transistors: electrical tests
Emitters:
  breakdown voltage, 45–6, 47–8
  current tests, 49–50, 50–1
  measurement of thermal resistance and circuit performance, 58–9, 59–60
  voltage tests, 49, 51–2
Environmental tests, 15–32
  barometric pressure, reduced (altitude operation), 15–16
  apparatus, 15–16
  details, 16

Environmental tests—*continued*
  procedure, 15
  dew point, 32
  high-temperature life (non-operating), 23
  immersion, 16–17
  insulation resistance, 17–19
    apparatus, 18–19
    details, 19
    factors affecting use, 18
    procedure, 19
  intermittent operation life test, 23–4
  moisture resistance, 19–22
    final measurements, 22
    procedure, 20–2
      initial conditioning, 20
      initial measurements, 20
      mounting, 20
      number of cycles, 20–1
      subcycle, 22
  salt-atmosphere corrosion test, 24–5
    apparatus, 24
    details, 25
    examination, 24–5
    procedure, 24
  salt-spray (corrosion) test, 25–9
    air supply, 26
    apparatus, 25–6
    atomisers, 26
    chamber, 26
    operating conditions, 28–9
      atomisation, 28
      details, 29
      length of test, 28–9
      measurements, 29
      temperature, 28
    preparation of specimens, 27
    procedure, 27–8
    salt solution, 26–7
  steady-state operation life test, 22–3
  temperature measurement: case and stud, 31–2
  thermal shock:
    glass strain, 31
    temperature cycling, 29–30
Environments, controlled, 14
External triggers, a.c. testing, 10
Extrapolated unity gain frequency, 77

Fatigue:
  leads, mechanical tests, 40
  vibration fatigue tests, 42
Field-effect transistors, tests, 114–24
  breakdown voltage, drain to source, 116
  breakdown voltage, gate to source, 114
  common-source, short-circuit:
    forward transadmittance, 122–3
    input admittance, 123–4
    input capacitance, 120
    output admittance, 121–2
  drain current, 117–18
  drain reverse current, 118
  drain-to-source, 'on'-state resistance, 119–20
  drain-to-source, 'on'-state voltage, 115
  gate-to-source, voltage or current, 115
  gate reverse current, 116–17
  static drain-to-source, 'on'-state resistance, 118–19
Figure of merit, microwave diode tests, 98–9
Floating potential test, transistors, 46–7
Flux, solderability tests, 35
  application, 36
  skimming, 37
Forward:
  current:
    general diode tests, 82
    transfer ratios:
      high- and low-frequency tests, 65–6, 78–9, 79–80
      transistors, 53–4
  impedance, general diode tests, 91
  leakage current, thyristors, 105–6
  recovery time, general diode tests, 83–4
  transadmittance, field-effect transistors, 122–3
  transfer, admittance, high- and low-frequency tests, 68
  voltage:
    general diode tests, 82
    thyristors, instantaneous forward voltage drop, 108
Frequency, high- and low-frequency tests, 64–80
  capacitance:
    depletion-layer, 73–4
    diffusion, 72–3
    direct inter-terminal, 72
    input, 71
    open-circuit output, 70–1
  extrapolated unity gain frequency, 77
  noise figure, 74
  pulse response, 74–6
  small-signal:
    open-circuit admittance, 65
    open-circuit reverse-voltage transfer ratio, 66–7

Frequency—*continued*
    power gain, 76–7
    real part of small-signal short-circuit input impedance, 77–8
    reverse-transfer admittance, 69
    short-circuit forward-current, transfer ratio, 65–6, 78–9
    short-circuit forward-transfer admittance, 68
    short-circuit input admittance, 67–8
    short-circuit input impedance, 64

Gain:
    extrapolated unity gain frequency test, 77
    small-signal power gain test, 76–7
Gates:
    field-effect transistors:
        breakdown voltage, gate-to-source test, 114
        gate-to-source voltage of current test, 115
        reverse current test, 116–17
    thyristors, gate triggering signal test, 107
Glass strain, thermal shock, 31

Heat:
    See also Thermal shock
    soldering, mechanical tests, 38
Heterodyne method, microwave diode conversion loss test, 96
Hex-base devices, case temperature measurements, 31–2
High-frequency tests *See* Frequency: high- and low-frequency tests
High-temperature, life (non-operating) test, 23
Holding current, thyristors, 105
Humidity, environmental tests, 19–22

Immersion tests, environmental testing, 16–17
Impedance:
    general diode tests:
        breakdown, 89–90
        forward, 91
    input, high- and low-frequency tests, 64, 77–8
    microwave diode tests:
        intermediate frequency, 99–101
        video, 102–3
Inaccessibility, as a factor of testing, 4

Incremental method, microwave diode conversion loss test, 95–6
Inductance, tunnel diode:
    negative, 111–12
    series, 111
Initial:
    conditioning, moisture resistance tests, 20
    measurements, moisture resistance tests, 20
Input:
    admittance:
        field effect transistors, 123–4
        high- and low-frequency tests, 67–8
    capacitance:
        field effect transistors, 120
        high- and low-frequency tests, 71
    impedance, high- and low-frequency tests, 64, 77–8
    resistance, transistors, 54–5
    small-signal power gain test, 76–7
Instantaneous forward voltage drop, 108
Insulation resistance, environmental tests, 17–19
Intermediate frequency impedance, microwave diode tests, 99–101
Intermittent operation, life test, 23–4
Inter-terminal capacitance, high- and low-frequency tests, 72

Johnson noise, 98
Junction capacitance, tunnel diode test, 109

Leads:
    bending stress test, 40–1
    fatigue, mechanical tests, 40
    order of connection, 3
    tabulation, intermittent operation life test, 24
    terminal strength tests, 38–9
Leakage:
    current, environmental tests, 17–19
    thyristors:
        forward leakage current, 105–6
        reverse leakage current, 106
Life tests:
    high-temperature (non-operating), 23
    intermittent operation, 23–4
    steady-state operation, 22–3
Loading effects:
    digital voltmeters, 6
    multimeters, 5–6
    oscilloscopes, 6

Loading effects—*continued*
  problems in testing, 6–8
  reduction by provision of monitoring points, 6
  supply arrangements, 8–9
Low-frequency *See* Frequency: high- and low-frequency tests
Lugs, solderability, 36, 37

Measurement/s:
  See also Meters
  salt-spray test, 29
  thermal shock, 30
Mechanical tests, 33–43
  bending stress, 40–1
  constant acceleration, 33
  lead fatigue, 40
  shock, 33–4
    simulated drop test, 34
  solderability, 34–7
    apparatus, 35
    materials, 35–6
    procedure, 36–7
  soldering heat, 38
  stud torque, 39–40
  terminal strength, 38–9
    lead or terminal torque, 38–9
    tension, 38
  vibration:
    fatigue, 42
    noise, 42–3
    variable frequency, 42
    monitored, 41
Megohm-meter, insulation resistance, 18
Merit, figure of, microwave diode tests, 98–9
Meters:
  See also Multimeters
  a.c. testing, 10
  accuracy, 1
  megohm-meters, 18
Microwave diodes, 95–104
  burnout tests, 103–4
  conversion loss test, 95–7
    heterodyne method, 96
    incremental method, 95–6
    modulation method, 96–7
    over-all noise method, 97
  detector power efficiency test, 97–8
  intermediate frequency impedance test, 99–101
  output noise ratio test, 101–2
  video impedance test, 102–3
Modulation method, microwave diode conversion loss test, 96–7

Moisture resistance, 19–22
Monitor points, reduction of loading effect, 6
Mounting:
  case temperature measurements, 32
  moisture resistance tests, 20
Multimeters, loading effects, 5–6

Negative inductance, tunnel diode tests, 111–12
Noise:
  figure:
    high- and low-frequency tests, 74
    microwave diode conversion loss test, 97
  Johnson noise, 98
  output, microwave diode tests, 101–2
  vibration noise test, 42–3
Non-destructive testing, 14

Open-circuits:
  See also Frequency: high- and low-frequency tests
  See also Tunnel diodes, tests
  output capacitance test, 70–1
Orientation, 2
Oscilloscopes:
  a.c. testing, 9–10
  loading effects, 6
Output:
  admittance:
    field-effect transistors, 121–2
    high- and low-frequency tests, 65, 69–70
  capacitance, high- and low-frequency tests, 70–1
  current, general diode tests (d.c.), 81–2
  noise, microwave diode tests, 101–2
  small-signal power gain test, 76–7

Potential, floating, electrical tests of transistors, 46–7
Power:
  gain, high- and low-frequency tests, 76–7
  supply:
    arrangements, 8–9
    problems in testing, 6–8
Preparation, of specimens, salt-spray test, 27
Pressurisation, environmental tests, 15–16
Prototypes, reliability of testing, 4

132 Semiconductor devices

Pulse/s:
  burnout:
    electrical tests of transistors, 45
    microwave diode tests, 103-4
  measurements, 3
  response:
    high- and low-frequency tests, 74-6
    thyristors, 106-7

Quality, variable capacitance diodes, 87-8
Quantitative reliability, 13

Real part, small-signal short-circuit input impedance test, 77-8
Recovery time, general diode tests:
  forward, 83-4
  reverse, 84-7
Rectification, efficiency, general diode tests, 88-9
Rectifiers, thermal resistance tests:
  controlled rectifiers, 93-4
  diodes, 93-4
Reliability:
  design for, 10-13
  of testing, 3-5
  'breadboard', 4
Resistance:
  field effect transistors:
    small-signal drain-to-source, 'on'-state resistance test, 119-20
    static drain-to-source, 'on'-state resistance, 118-19
  saturation, electrical tests of transistors, 52-3
  series, tunnel diode tests, 112-13
  static input, transistors, 54-5
  thermal *See* Thermal: resistance
Reverse:
  current:
    field-effect transistors:
      drains, 118
      gates, 116-17
    general diode tests, 82-3
    average, 89
  leakage current, thyristors, 106
  recovery time, general diode tests, 84-7
  transfer:
    admittance, high- and low-frequency tests, 69
    field-effect transistors, capacitance, 121

voltage:
  general diode tests, 82-3
  high- and low-frequency tests, 66-7
Rogue failures, elimination by accelerated life tests, 13-14

Salt:
  salt-atmosphere (corrosion) test, 24-5
  salt-spray (corrosion) test, 25-9
Saturation:
  current, general diode tests, 92-3
  voltage and resistance tests of transistors, 52-3
Sea-coast atmospheres, environmental tests, 24-9
Seals:
  bending stress test, 40-1
  environmental tests, 16-17
Semiconductors, transistors, electrical tests *See* Transistors: electrical tests
Series, tunnel diode tests:
  inductance, 111
  resistance, 112-13
Shock:
  test for, 33-4
  thermal *See* Thermal: shock
  capacitance, high- and low-frequency tests, 71, 74
Short-circuits:
  See Field effect transistors
  See Frequency: high- and low-frequency tests
  See Tunnel diode tests
Signal diodes, thermal resistance tests, 93-4
Simulated drop, shock test, 34
Small-signals:
  See Diodes: general tests
  See Field-effect transistors
  See Frequency: high- and low-frequency tests
  See Tunnel diodes, tests
Solder dip, 37
Solderability, mechanical tests, 34-7
  apparatus, 35
  materials, 35-6
  procedure, 36-7
    ageing, 36
    application of flux, 36
    application of standard solderable wire for lugs, tabs, 36
    evaluation of lugs, tabs, 37
    evaluation of solid wire terminations, 37

Solderability—*continued*
    examination of terminations, 36–7
    preparation of terminations, 36
    solder dip, 37
Soldering:
    heat, mechanical tests, 38
Spray, salt-spray corrosion test, 25–9
Stabilisers, voltage, 8–9
Static:
    characteristics, tunnel diode tests, 109–11
    drain, field-effect transistors, 118–19
    input, transistors, resistance test, 54–5
    transconductance, electrical tests of transistors, 55–6
Steady-state operation life test, 22–3
Step-stress testing, 12–13
Stored charge, general diode tests, 91–2
Strength, terminal, mechanical tests, 38–9
    lead or terminal torque, 38–9
    tension, 38–9
Stress bending, mechanical tests, 40–1
Stud torque, mechanical tests, 39–40
Surge current, general diode tests, 90
Switching time, tunnel diode tests, 113

Tabs, solderability, 36, 37
Temperature:
    See also Thermal: shock
    coefficient, breakdown, voltage test, general diodes, 90–1
    cycling, environmental tests, 29–30
    measurement, case and stud, environmental tests, 31–2
    permissible variation in environmental chambers, 1
    salt-spray test, 28
Tension, terminal strength test, 38
Terminals:
    direct inter-terminal capacitance test, 72
    floating potential test, 46–7
    strength, mechanical tests, 38–9
    lead or terminal torque, 38–9
    tension, 38
Terminations, examination and evaluation, solderability, 37
Test conditions, 1
    electrical measurements, 3
Testing, reliability, 3–5
Thermal:
    resistance:
        general diode tests, 93–4
        measurements, 57–63

        collector cut-off method, 57–8
        d.c. current gain continuous method, 62–3
        d.c. forward voltage drop, emitter base continuous method, 59–60
        forward voltage drop, collector-to-base diode method, 60–1
        forward voltage drop, emitter-to-base diode method, 58–9
        general, 62
        thermal response time test, 60
        thermal time constant test, 61–2
    reliability, 3–5
    shock:
        glass strain, environmental tests, 31
        temperature cycling, environmental tests, 29–30
Thermocycles, case temperature measurements, 32
Thyristors, 105–8
    forward leakage tests, 105–6
    gate triggering signal test, 107
    holding current test, 105
    instantaneous forward voltage drop, 108
    pulse response test, 106–7
    reverse leakage current test, 106
Time, thermal time constant test, 61–2
Torque:
    lead or terminal strength test, 38–9
    stud, mechanical tests, 39–40
Transadmittance, field-effect transistors, 122–3
Transconductance, static, electrical tests of transistors, 55–6
Transfer:
    ratios, forward-current, tests of transistors, 53–4
    See also Frequency: high- and low-frequency tests
    time, thermal shock test, 31
Transients, 3
Transistors:
    electrical tests, 44–56
        base-to-emitter voltage (saturated or non-saturated) test, 51–2
        breakdown voltage, collector-to-base, 44
        breakdown voltage, collector-to-emitter, 45–6
        burnout by pulsing, 45
        collector-to-base current test, 48
        collector-to-base voltage test, 50
        collector-to-emitter cut-off current test, 49–50

Transistors—*continued*
  collector-to-emitter voltage test, 49
  drift, 46
  emitter-to-base current test, 50–1
  emitter-to-base test, 47–8
  floating potential test, 46–7
  forward-current transfer ratio test, 53–4
  saturation voltage and resistance test, 52–3
  static input resistance test, 54–5
  static transconductance test, 55–6
  field-effect *See* Field-effect transistors
Triggering, gate triggering signal test, 107
Tropical degradation, environmental testing, 19–22
Tunnel diodes, tests, 109–13
  junction capacitance, 109
  negative inductance, 111–12
  series inductance, 111
  series resistance, 112–13
  static characteristics, 109–11
  switching time, 113

Unity gain, extrapolated unity gain frequency test, 77

Vibration:
  fatigue test, 42

noise test, 42–3
variable frequency test, 41–2
Video:
  crystals, figure of merit test, 98–9
  impedance, microwave diode tests, 102–3
Voltage:
  breakdown *See* Breakdown: voltage
  drop, measurement of thermal resistance and circuit performance, 58ff
  forward *See* Forward: voltage
  reverse *See* Reverse: voltage
  stabilisers, 8–9
  thyristors, rate of noise test, 108
  tests:
    field-effect transistors
      drain-to-source, 'on'-state, 115
      gate-to-source, 115
    transistors:
      base-to-emitter, 51–2
      collector-to-base, 50
      collector-to-emitter, 49
      saturation, 52–3
Voltmeters:
  analogue, a.c. testing, 10
  digital:
    a.c. testing, 10
    loading effects, 6

Welds, terminal strength test, 38
Wire, wrapping, solderability tests, 35–6
  application, 36